SOFT-SWITCHING PWM FULL-BRIDGE CONVERTERS

SOFT-SWITCHING PWM FULL-BRIDGE CONVERTERS

TOPOLOGIES, CONTROL, AND DESIGN

Xinbo Ruan
Nanjing University of Aeronautics and Astronautics, China

WILEY

SCIENCE PRESS

This edition first published 2014

Published by John Wiley & Sons Singapore Pte. Ltd., 1 Fusionopolis Walk, #07-01 Solaris South Tower, Singapore 138628, under exclusive license by Science Press in all media throughout the world excluding Mainland China and excluding Simplified and Traditional Chinese languages.

For details of our global editorial offices, for customer services and for information about how to apply for permission to reuse the copyright material in this book please see our website at www.wiley.com.

Library of Congress Cataloging-in-Publication Data has been applied for.

A catalogue record for this book is available from the British Library.

ISBN: 978-1-118-70220-8

Typeset in 11/13pt Times by Laserwords Private Limited, Chennai, India

1 2014

Contents

About the Author

Xinbo Ruan was born in Hubei Province, China, in 1970. He received a BS and a PhD in electrical engineering from Nanjing University of Aeronautics and Astronautics (NUAA), Nanjing, China, in 1991 and 1996, respectively.

In 1996, he joined the Faculty of Electrical Engineering Teaching and Research Division, NUAA, where he became a Professor in the College of Automation Engineering in 2002 and engaged in teaching and research in the field of power electronics. From August to October 2007, he was a Research Fellow in the Department of Electronic and Information Engineering, Hong Kong Polytechnic University, Hong Kong, China. Since March 2008, he has been also with the School of Electrical and Electronic Engineering, Huazhong University of Science and Technology, China. He is a Guest Professor at Beijing Jiaotong University, Beijing, China, Hefei University of Technology, Hefei, China, and Wuhan University, Wuhan, China. He is the author or co-author of 4 books and more than 100 technical papers published in journals and conferences. His main research interests include soft-switching dc–dc converters, soft-switching inverters, power-factor-correction converters, modeling of converters, power electronics system integration, and renewable energy generation systems.

Dr. Ruan was a recipient of the Delta Scholarship from the Delta Environment and Education Fund in 2003 and became Special Appointed Professor of the Chang Jiang Scholars Program by the Ministry of Education, China, in 2007. From 2005 to 2013, he served as Vice President of the China Power Supply Society, and since 2008 he has been a member of the Technical Committee on Renewable Energy Systems within the IEEE Industrial Electronics Society. He has been an Associate Editor for the IEEE *Transactions on Industrial Electronics* and the IEEE *Journal of Emerging and Selected Topics on Power Electronics* since 2011 and 2013, respectively. He is a Senior Member of the IEEE Power Electronics Society and the IEEE Industrial Electronics Society.

Preface

During the past 2 decades, power electronics technology has been rapidly developing, and power electronics converters have found various applications in information technology, home appliances, electrified transportation, renewable-energy-generating systems, and electrical power systems. High efficiency, high power density, and high reliability are continual requirements for power electronics converters, in addition to electrical specifications and performance. Soft-switching technology is a promising approach to meeting these, since it allows the power devices to realize zero-voltage switching (ZVS) or zero-current switching (ZCS) and significantly reduces the switching loss.

Unlike forward, push–pull, and half-bridge converters, the full-bridge converter, which has four power switches, is suitable for medium-to-high power conversions. This book aims to systematically describe the soft-switching techniques for isolated buck-derived full-bridge converters, including topologies, control, and design, and will reveal the relationships among the various topologies and pulse-width modulation (PWM) strategies. For brevity of illustration, the isolated buck-derived full-bridge converter is simply called a full-bridge converter hereinafter.

This book includes eight chapters.

In Chapter 1, the single-switch forward, dual-switch forward, push–pull, half-bridge, and full-bridge converters are derived, revealing the relationships among them. The derivation of full-wave rectifier, full-bridge rectifier, and current-doubler rectifier circuits is also presented, for the same reason. Then, the basic operating principle of full-bridge converters with various output rectifier circuits is presented, which provides the basis for the descriptions in the subsequent chapters.

In Chapter 2, a family of PWM strategies for the full-bridge converters is presented, and they are categorized into two types. It is pointed out that the two diagonal switches should be turned off at different times in order to allow the full-bridge converters to realize soft switching. Thus, the concept of leading and lagging legs is introduced: the leading leg can only realize ZVS, but can do so easily, whereas the lagging leg can realize either ZVS or ZCS. Thus, the soft-switching PWM full-bridge converters are classified into two kinds, namely ZVS and zero-voltage-and-zero-current switching (ZVZCS).

In Chapter 3, the topologies of the ZVS PWM full-bridge converters are discussed and suitable modulation strategies from the family of PWM strategies are pointed out. The phase-shift-controlled ZVS PWM full-bridge converter is taken as an example to illustrate the operating principle. The conditions required for the leading and lagging legs to achieve ZVS are discussed, and it is pointed out that the leading leg can do so easily, while for the lagging leg it is more difficult. The duty cycle loss resulting from the resonant inductor is also analyzed.

Chapter 4 first describes the current-enhancement principle, in which an auxiliary-current-source network is added to help the resonant inductor achieve ZVS for the lagging leg in a wide load range. Then, a simple auxiliary-current-source network is introduced to the ZVS PWM full-bridge converter, whose operating principle is analyzed. This converter achieves ZVS for all power switches in a wide load range, and its duty cycle loss is reduced. Finally, other auxiliary-current-source networks capable of increasing the ZVS load range are discussed.

In Chapter 5, the zero-state current-reset scheme of the PWM full-bridge converter is discussed, and several methods for resetting the primary current in the zero state are presented. Thus, several ZVZCS PWM full-bridge converters are derived. The ZVZCS PWM full-bridge converters, in which a diode is introduced in series with each lagging-leg switch and a blocking capacitor is added in series with the primary winding of the transformer, is taken as an example to illustrate the operating principle. This converter can achieve ZVS for the leading leg and ZCS for the lagging leg in a wide load range.

These chapters all focus on the achievement of soft switching for the power switches of the full-bridge converters. Chapters 6–8 are dedicated to depressing the voltage oscillation across the output rectifier diodes of the ZVS PWM full-bridge converters.

In Chapter 6, the cause of voltage oscillation across the output rectifier diodes is explained, and methods of depressing the voltage oscillation are discussed. The operating principle of the ZVS PWM full-bridge converters with two clamping diodes is analyzed; these introduced clamping diodes can eliminate voltage oscillation across the output rectifier diodes.

Based on the ZVS PWM full-bridge converters with two clamping diodes, Chapter 7 presents a clamping diode current-reset scheme based on the addition of an auxiliary transformer. This allows the clamping diode current to be reduced to zero rapidly over the full load range. As a consequence, the conduction loss is reduced and the serious reverse recovery of the clamping diodes at light load is avoided. This greatly improves the efficiency and reliability of the full-bridge converter.

Chapter 8 presents an improved ZVS PWM full-bridge converter with a current-doubler rectifier. This converter can achieve ZVS for both the leading and the lagging legs over the whole load range, and the rectifier diodes commute naturally without voltage oscillation. The design of the output filter inductance and blocking capacitor is discussed.

This book is written for Master and PhD students and teachers majoring in power electronics and for power supply design engineers. It also provides an excellent opportunity for senior undergraduate students majoring in electrical engineering and automation engineering to become familiar with the latest developments in the field.

Acknowledgment

This monograph systematically summarizes the research work on soft-switching full-bridge dc–dc converters that I, my PhD supervisor, and my PhD students have continuously carryied out over the past nearly 20 years.

First of all, I would like to thank Prof. Yangguang Yan, my PhD supervisor at Nanjing University of Aeronautics and Astronautics (NUAA), for his encouragement when I chose this topic as my PhD dissertation. Every discussion with him brought me new ideas, which made this research work more and more extensive.

I wish to thank my former PhD students, Jianggang Wang, Fuxin Liu, Wu Chen, and Qianhong Chen for their great efforts and their significant contribution to this book. I also wish to thank my students Xin Zhang and Yalong Li for building a new prototype of a zero-voltage-switching (ZVS) pulse-width modulation (PWM) full-bridge converter and updating the experimental results.

Special thanks is given to Prof. Chi. K. Tse at Hong Kong Polytechnic University for his great efforts in polishing the writing of this book, the readability of which is greatly improved.

I also wish to thank Prof. Daohong Ding at NUAA and Prof. Xuansan Cai at Tsinghua University for their valuable suggestions for the Chinese version of this book. Profs Ding and Cai have passed away, and I would like to dedicate this book as a memorial to them.

It has been a great pleasure to work with the staff of John Wiley & Sons, Ltd., Jayashree Saishankar, Project Manager and Science Press, China. The support and help from Minxin Hou (the Commissioning Editor) and Clarissa Lim (the Project Editor) are greatly appreciated.

Xinbo Ruan
Aero-Power Sci-Tech Center, College of Automation Engineering,
Nanjing University of Aeronautics and Astronautics,
Nanjing, China
October 2013

List of Abbreviations

ac	alternating current
CCM	continuous current mode
CDR	current-doubler rectifier
dc	direct current
DCM	discontinuous current mode
GaN	gallium nitride
IGBT	insulated-gate bipolar transistors
MOSFET	metal-oxide-semiconductor field-effect transistors
MTBF	mean time between failure
PWM	pulse-width modulation
RC	resistor–capacitor
RCD	resistor–capacitor–diode
RDCLI	resonant dc link inverters
SEPIC	single-ended primary inductor converter
SiC	silicon carbide
ZCS	zero-current switching
ZCT	zero-current transition
ZVS	zero-voltage switching
ZVT	zero-voltage transition
ZVZCS	zero-voltage-and-zero-current switching

1

Topologies and Operating Principles of Basic Full-Bridge Converters

1.1 Introduction

1.1.1 Development Trends of Power Electronics Technology

High-frequency switching is one of the most important features of power electronics technology, enabling power electronics converters to meet the required specifications and performance. High-frequency power devices and components, including magnetic devices and capacitors, are the basic elements of high-frequency power electronics. Metal-oxide-semiconductor field-effect transistors (MOSFETs) and insulated-gate bipolar transistors (IGBTs) have become the dominant choices for the implementation of power switches, and MOSFETs with low gate charges and low junction capacitors further boost the development of high-frequency power electronics. Recently, there has been significant progress in the development of silicon carbide (SiC)-based power devices, including SiC diodes [1], SiC MOSFETs, and SiC IGBTs [2], and SiC-based commercial products have become strong competitors to Si-based fast-recovery diodes and MOSFETs in medium power conversion applications. Gallium nitride (GaN) power devices have drawn attention for achieving ultra-fast switching. Furthermore, recent progress in the development of amorphous, microcrystalline cores and high-frequency ferrites has been significant.

Circuit topology is another important aspect of high-frequency power electronics. Switching losses in switching devices have been drastically reduced through the systematic development of resonant converters [3, 4], quasi-resonant converters [5], and multi-resonant converters [5, 6], zero-voltage-switching (ZVS) pulse-width modulation (PWM) and zero-current-switching (ZCS) PWM converters [7, 8],

Soft-Switching PWM Full-Bridge Converters: Topologies, Control, and Design, First Edition. Xinbo Ruan.

zero-voltage-transition (ZVT) and zero-current-transition (ZCT) converters [9, 10], and resonant dc link inverters (RDCLIs) [11] that partially or fully achieve ZVS and ZCS. As a result, the operating switching frequency has increased by an order of magnitude and more. High-frequency switching is also the key contributor to miniaturization and modularization, due to the significant gain in efficiency of power conversion that it can offer, in addition to the high insulation and high thermal conductivity of the structure employed. Therefore, the success of high-frequency power electronics can be attributed to the advent of high-frequency power devices and components, soft-switching technologies, mechanical structures, materials, and related technologies.

1.1.2 Classification and Requirements of Power Electronics Converters

Power electronics converters are a family of electrical circuits that convert electrical energy from one level of voltage/current/frequency to another using semiconductor devices, passive components, and advanced control methods. According to the form of conversion, power electronics converters can be classified into four different types [12]: (i) dc–dc converters, which convert a dc input voltage into a dc output voltage of a different magnitude and possibly opposite polarity, or with galvanic isolation of the input and output ground references; (ii) dc–ac inverters, which transform a dc input voltage into an ac output voltage of controllable magnitude and frequency; (iii) ac–dc rectifiers, which convert an ac input voltage into a dc output voltage, and are capable of controlling the dc output voltage and/or ac input current waveform; and (iv) ac–ac cycloconverters, which convert an ac input voltage into an ac output voltage of controllable magnitude and frequency. These four kinds of power electronics converter can be unidirectional or bidirectional. The unidirectional ones can only convert electrical power from a defined input terminal to a defined output terminal, while the bidirectional ones can convert electrical power in either direction between two defined terminals.

The primary objective of power electronics converters is to meet the corresponding electrical specifications and regulatory requirements. While meeting the electrical specifications, a power electronics converter should idealy achieve high efficiency, high power density, high reliability, and low cost. High efficiency not only leads to energy saving but also results in a reduced heat dissipation requirement. High power density means compact size at the required output power, which is very important for aerospace applications and compact portable appliances. Power converters of high reliability are more competitive in the commercial market and can be essential for critical applications that require operation under adverse working conditions and long mean-time-between-failure (MTBF). Moreover, the overall cost is a key factor for commercial power supply applications. Power electronics converters are often required to have good maintainability, including reduced technical requirements for repair personnel and shortened repair times.

1.1.3 Classification and Characterization of dc–dc Converters

Dc–dc converters are an important kind of power electronics converter. With the development of power electronics technology, computer science and technology, and information technology, dc–dc converter-based switching-mode power supplies have been widely used. The dc–dc converter is thus the key basic building block in power electronics and has attracted a great deal of research attention in the past few decades.

According to the presence of galvanic isolation between input and output, dc–dc converters can be divided into two classes: non-isolated and isolated. Basic non-isolated dc–dc converters include buck, boost, buck–boost, Cuk, Zeta, and single-ended primary inductor converter (SEPIC) converters. Other examples include the dual-switch buck–boost converter, the full-bridge converter, and so on.

Isolated dc–dc converters are derived from non-isolated dc–dc converters by incorporating transformers and output rectifier circuits. Isolated buck-derived converters include forward, push–pull, half-bridge, and full-bridge. The forward converters can be single-switch or dual-switch versions. Isolated boost-derived converters include push–pull, half-bridge, and full-bridge versions. Isolated buck-boost converters are flyback converters, which can also be single-switch or dual-switch versions. Cuk, Zata, and SEPIC converters also have isolated versions.

With the power devices having the same voltage and current ratings, the output power of the dc–dc converter is proportional to the number of power switches. Thus, the isolated dc–dc converter with two power switches (dual-switch forward, push–pull, and half-bridge) can handle twice as much power as an isolated dc–dc converter with only one power switch, such as the single-switch forward converter, and only half as much power as a converter with four power switches, such as the full-bridge converter. Thus, the full-bridge converter can handle the largest power among all the isolated buck-derived converters, and it has been widely used in high-input-voltage and medium- to high-power conversion applications.

Resonant converters, quasi-resonant converters, and multi-resonant converters can achieve ZVS or ZCS for power switches without using additional auxiliary power switches. However, such soft-switching converters are different from the conventional PWM converters and they have the disadvantages of high voltage/current stress, large circulating energy, and variable switching frequency. For ZVS-PWM and ZCS-PWM converters, the operating frequency is constant and additional auxiliary power switches are needed. Also, the voltage/current stress of both the main and auxiliary power switches is relatively high. For ZVT and ZCT converters, the operating frequency is also fixed. However, the additional auxiliary power switches are used only for achieving ZVT or ZCT for the main power switches and do not contribute to power processing. Among the isolated buck-derived converters, the converters with two or four power switches can achieve ZVT or ZCT without additional auxiliary switches, provided an appropriate control scheme is employed. Such simplicity makes them popular choices in practical applications. This book intends to describe systematically the soft-switching techniques for isolated buck-derived full-bridge converters. For brevity of illustration, the isolated buck-derived full-bridge converter is simply called a "full-bridge converter."

1.2 Isolated Buck-Derived Converters

In order to give some insights into the characteristics of the isolated buck-derived converters and to reveal the relationships among them, this section begins by deriving the forward converter from the basic buck converter. It then goes on to derive the dual-switch forward converter, push–pull converter, half-bridge converter, and full-bridge converter.

1.2.1 Forward Converter

1.2.1.1 Derivation of a Single-Switch Forward Converter

The buck converter is the most basic of the dc–dc converters. It is shown in Figure 1.1a, where V_{in} is the input dc voltage, Q is the power switch, D_{FW} is the freewheeling diode, and L_f and C_f are the output filter inductor and output filter capacitor, respectively. In order to achieve galvanic isolation, the transformer T_r is inserted between Q and D_{FW}, as shown in Figure 1.1b. The primary and secondary winding turns of T_r are N_p and N_s, respectively, and the corresponding turns ratio is $K = N_p/N_s$. When Q conducts, the input voltage V_{in} is applied on the transformer primary winding. The transformer is thus magnetized and the magnetizing flux ϕ_m increases.

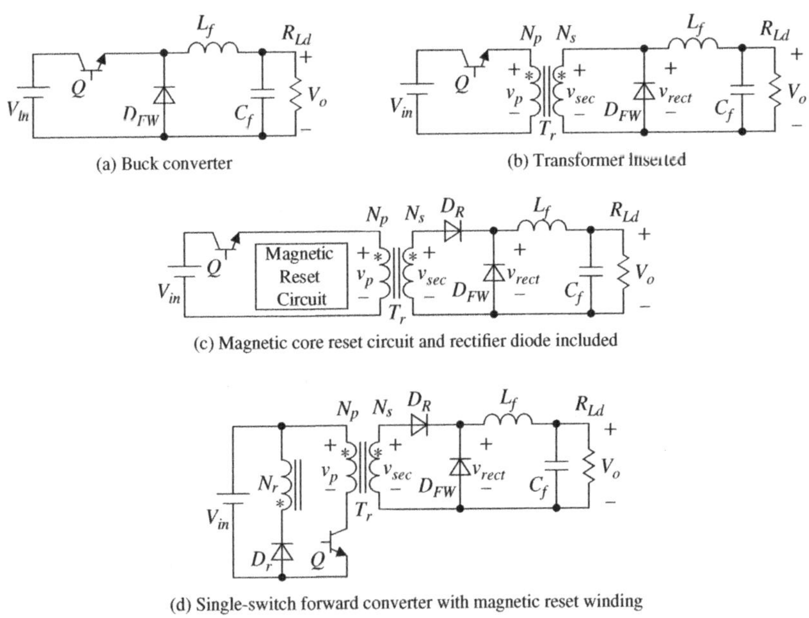

(a) Buck converter

(b) Transformer inserted

(c) Magnetic core reset circuit and rectifier diode included

(d) Single-switch forward converter with magnetic reset winding

Figure 1.1 Derivation of the single-switch forward converter

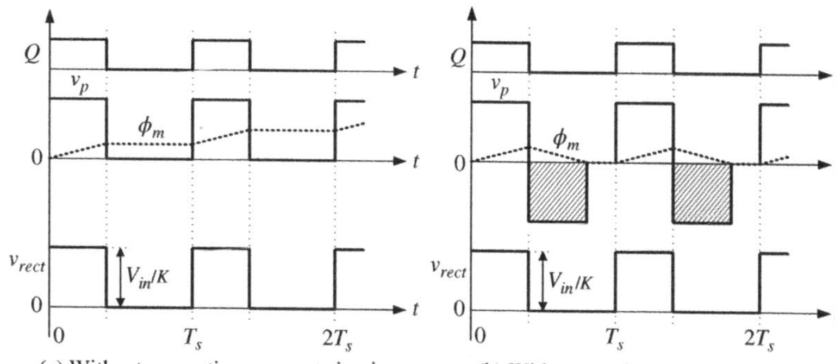

Figure 1.2 Waveforms of the primary voltage and magnetizing flux of the transformer

When Q is turned off, the filter inductor current is freewheeling through diode D_{FW}. The conducting diode short-circuits the transformer secondary winding, forcing the primary winding voltage to zero and thus keeping ϕ_m unchanged. Therefore, in a switching period, ϕ_m has a net increase, and as time elapses, the magnetic core will saturate, leading to destruction of the power switches. The waveforms of the primary voltage v_p and magnetizing flux ϕ_m of the transformer are sketched in Figure 1.2a.

In order to avoid saturating the transformer, the magnetizing flux should be reset before the end of each switching period. Thus, a magnetic core reset circuit is mandatory. This circuit applies a negative voltage across the primary winding of the transformer when Q is turned off, as shown in the shaded area in Figure 1.2b. However, this negative voltage will be reflected to the secondary winding and will force the freewheeling diode D_{FW} to conduct, shorting the secondary winding. In order to avoid short-circuiting the transformer, a diode D_R can be inserted in series with the secondary winding, as shown in Figure 1.1c. Thus, a magnetic core reset circuit can be formed by a reset winding N_r and a reset diode D_r. Exchanging the positions of power switch Q and the transformer primary winding, the basic single-switch forward converter is formed, as shown in Figure 1.1d. In practical applications, the turns of the reset winding and primary winding are usually designed to be equal, and thus the voltage stress of the power switch is $2V_{in}$ and its maximum duty cycle is limited to 0.5, in order to achieve magnetic core reset of the transformer.

1.2.1.2 Derivation of a Dual-Switch Forward Converter

As the power switch sustains twice the input voltage, the single-switch forward converter is suitable for low-input-voltage applications. For high-input-voltage applications, it may be difficult to find an appropriate power switch with such a high voltage rating. For example, with a single-phase input ac voltage of 220 V ± 20%, the rectified input voltage with power factor correction is 380 V. The voltage stress of the power switch is 760 V and a power switch with voltage rating of 1000 V is required. MOSFETs with such a high voltage rating have poor performance at high frequencies, and the drain-source conduction resistor $R_{ds(on)}$ is relatively large. Although IGBTs can be adopted, the switching frequency is limited to tens of kilohertz due to the presence of the current tail. At high frequencies, the turn-off loss will be relatively high, causing degradation in efficiency.

For the sake of employing the available power switches with better performance, it is desirable to reduce the voltage stress of the power switch. As indicated before, when the turns of primary winding and reset winding are equal, the power switch of a single-switch forward converter must withstand twice the input voltage. If power switch Q is replaced by two power switches Q_1 and Q_2, as shown in Figure 1.3a, the voltage stress of Q_1 and Q_2 will be the input voltage V_{in}. Exchanging the positions of Q_1 and the primary winding leads to the circuit shown in Figure 1.3b. In order to ensure the voltage stress of Q_1 and Q_2 is V_{in}, a diode D_2 is inserted between point A and the negative rail and a diode D_1 between point B and the positive rail, as shown in Figure 1.3c. When both Q_1 and Q_2 are turned off, the transformer is magnetically reset through the reset winding N_r and the transformer primary voltage v_{AB} is $-V_{in}$. In fact, the transformer can be magnetically reset through the primary winding via D_1 and D_2. Therefore, the path constituting reset winding N_r and reset diode D_r is redundant and can be removed. The simplified circuit is redrawn in Figure 1.3d, giving the well-known dual-switch forward converter. In this circuit, the voltage stress of the

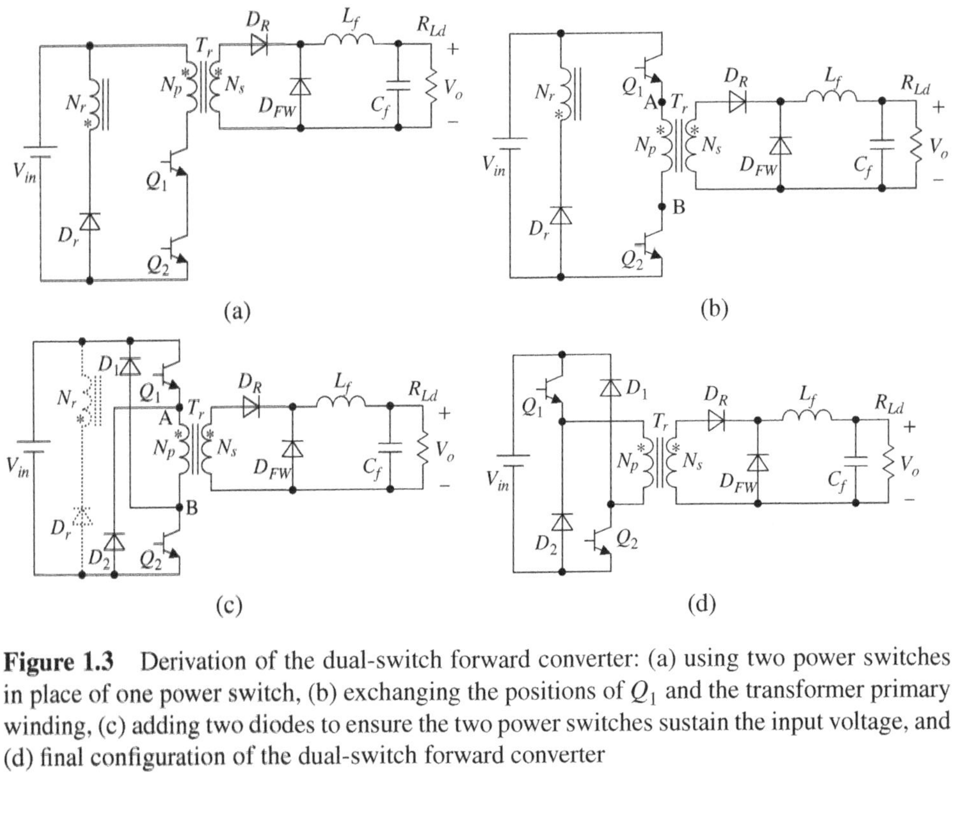

Figure 1.3 Derivation of the dual-switch forward converter: (a) using two power switches in place of one power switch, (b) exchanging the positions of Q_1 and the transformer primary winding, (c) adding two diodes to ensure the two power switches sustain the input voltage, and (d) final configuration of the dual-switch forward converter

two power switches is V_{in}, which is half that of the single-switch forward converter. Moreover, D_1 and D_2 are the reset diodes, and also provide the path for regeneration of the leakage inductor energy to the input voltage when Q_1 and Q_2 are turned off.

1.2.2 Push–Pull Converter

For the single-switch forward converter, when the primary winding and reset winding have the same number of turns, the duty cycle should be bounded below 0.5 to ensure magnetic core reset. Therefore, the magnitude of the secondary rectified voltage should be larger than twice the output voltage. This secondary rectified voltage contains a large amount of harmonics, necessitating the use of a large output inductor. In order to reduce the magnitude of the secondary rectified voltage and hence the output filter inductor, we can use two single-switch forward converters connected in parallel at the input side and the secondary rectifier, sharing the freewheeling diode and the output filter, as shown in Figure 1.4a. Note that the two forward converters

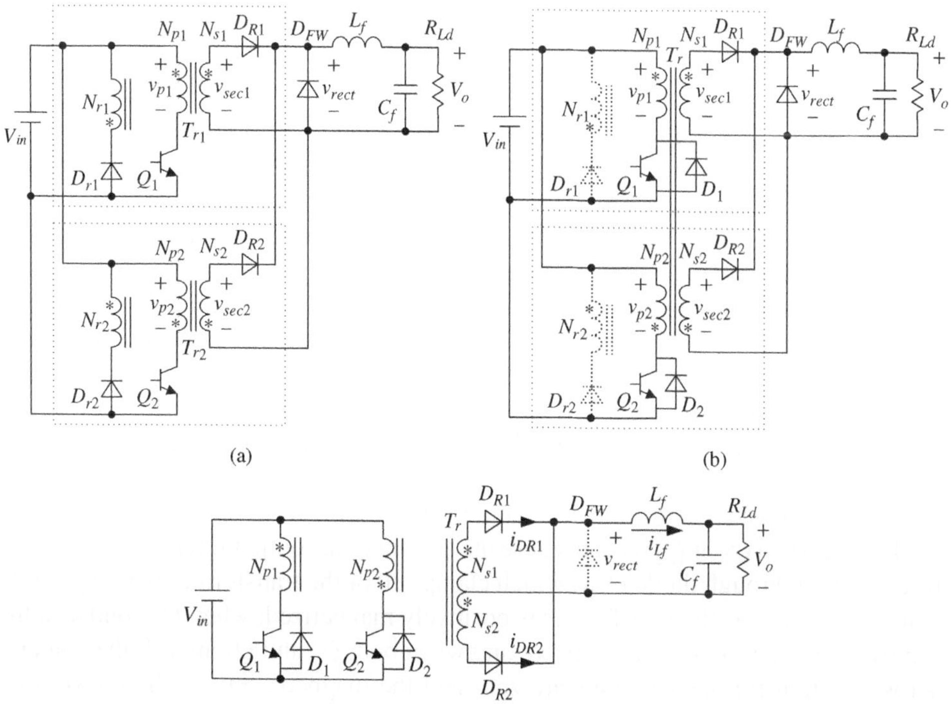

(a) (b) (c)

Figure 1.4 Derivation of the push–pull converter: (a) two single-switch forward converters connected in parallel at the input side and the secondary rectifier, sharing the freewheeling diode and the output filter, (b) the two transformers sharing a magnetic core, and (c) final configuration of the push–pull converter

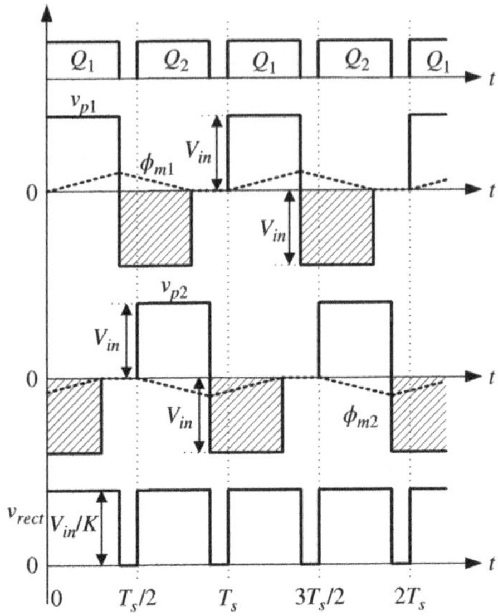

Figure 1.5 Key waveforms of two interleaved forward converters

are required to operate in an interleaving manner; that is, power switches Q_1 and Q_2 operate at the same switching frequency with a time difference of half the switching period, as shown in Figure 1.5.

Referring to Figure 1.4a, if we let the two transformers share a magnetic core and connect an antiparallel diode D_2 across Q_2, as shown in Figure 1.4b, the primary winding N_{p2} and diode D_2 can provide the magnetic core reset function for the transformer when Q_1 is turned off. Likewise, by connecting an antiparallel diode D_1 across Q_1, the transformer core can be magnetically reset through the primary winding N_{p1} and D_1 when Q_2 is off. This makes the two core reset circuits redundant and they can be removed. The simplified circuit is re-sketched in Figure 1.4c, showing what is usually referred to as the push–pull converter.

It should be noted that the transformer of the push–pull converter cannot be magnetically reset when Q_1 or Q_2 is off because the current of the output filter inductor will be freewheeling through diode D_{FW}, which clamps all of the transformer winding at 0 V. When Q_1 conducts, the transformer is positively magnetized; when Q_2 conducts, the transformer is negatively magnetized; and when both Q_1 and Q_2 are off, the voltages across the transformer windings are zero and the magnetic flux of the transformer remains unchanged. Figure 1.6 shows the key waveforms of the push–pull converter. For the forward converter, including the single-switch and dual-switch versions, the magnetizing current can only flow in a single direction, as when the magnetizing current decays to zero, it cannot flow in the reverse direction. For the push–pull converter,

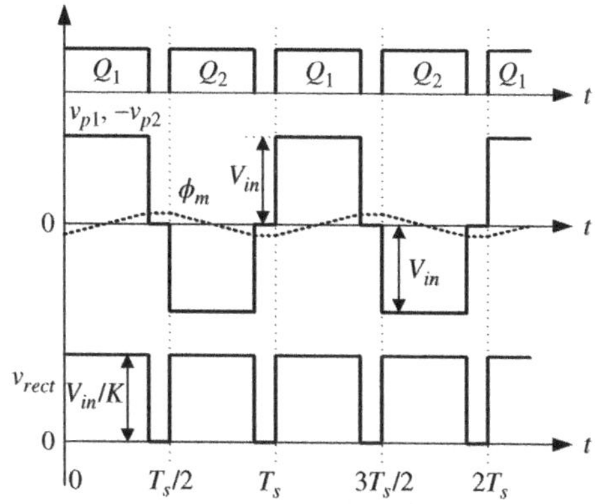

Figure 1.6 Key waveforms of the push–pull converter

the magnetizing current of the transformer flows bidirectionally. If the flux swing is constrained by the core loss rather than by the saturation flux density, the utilization of the transformer of the forward converter is the same as that of the push–pull converter.

As in the single-switch forward converter, the two switches of the push–pull converter must withstand a voltage of $2V_{in}$. Since the push–pull converter is equivalent to the interleaved parallel connection of two single-switch forward converters, the ripple frequency of the secondary rectified voltage v_{rect} is twice the switching frequency and its duty cycle can reach unity, as shown in Figure 1.6. With the same input and output voltages, the required magnitude of v_{rect} of the push–pull converter is only half that of the forward converter. Thus, the primary-to-secondary-winding-turns ratio of the push–pull converter is twice that of the forward converter.

Referring to Figure 1.4c, when both switches are off, the output filter inductor current can freewheel through D_{FW} or flow through the two secondary windings via rectifier diodes D_{R1} and D_{R2}. Therefore, D_{FW} is redundant and can be removed. When D_{FW} is removed, we get:

$$i_{DR1} + i_{DR2} = i_{Lf} \tag{1.1}$$

For an ideal transformer, the magnetizing current is zero; that is:

$$i_{DR1} - i_{DR2} = 0 \tag{1.2}$$

According to Equations 1.1 and 1.2, we have:

$$i_{DR1} = i_{DR2} = i_{Lf}/2 \tag{1.3}$$

Hence, when both the power switches are off, the output filter inductor current is shared by the two rectifier diodes D_{R1} and D_{R2}, with D_{FW} removed.

1.2.3 Half-Bridge Converter

Figure 1.7a shows two single-switch forward converters connected in series at the input side and in parallel at the secondary rectifiers, sharing the freewheeling diode and the output filter, where C_{d1} and C_{d2} are two input dividing capacitors. The values of the two capacitors are equal and quite large, and the voltage across each is $V_{in}/2$. Power switches Q_1 and Q_2 operate at the same switching frequency with a time difference of half the switching period $T_s/2$. If the two transformers share a magnetic core and an antiparallel diode D_2 is connected across Q_2, the transformer core can be magnetically reset through primary winding N_{p2} and D_2 when Q_1 is off. Likewise, inserting an antiparallel diode D_1 across Q_1, the transformer core can be magnetically reset through primary winding N_{p1} and D_1 when Q_2 is off. Thus, the two magnetic core reset circuits are redundant and can be removed, as shown in Figure 1.7b. Exchanging the positions of $Q_1(D_1)$ and primary winding N_{p1} leads to the circuit shown in Figure 1.7c. There is current flowing through each of the two primary windings when

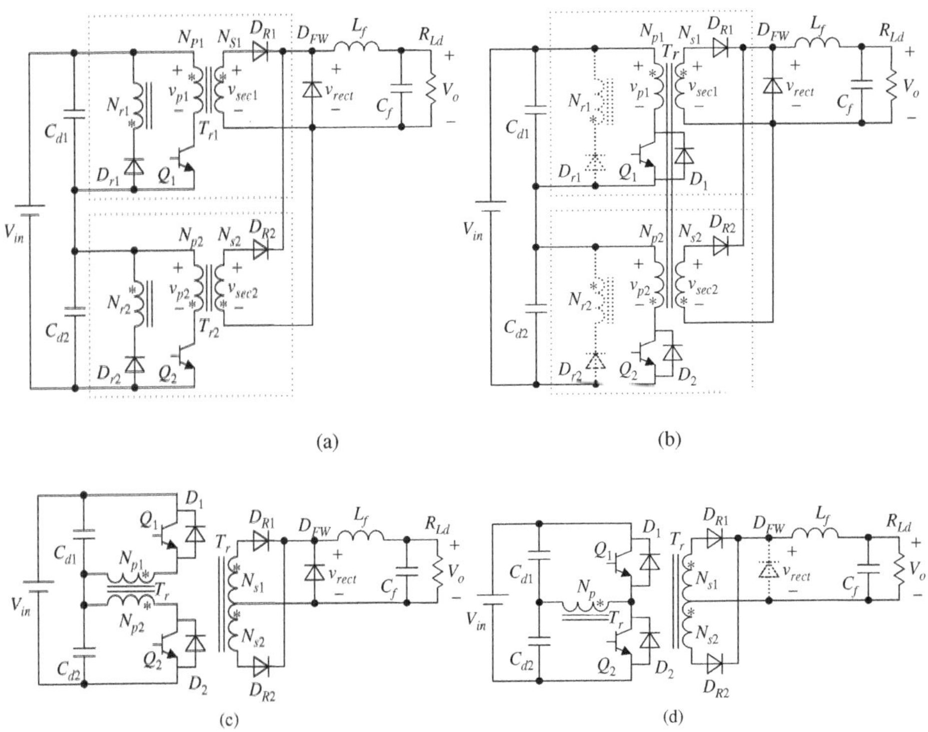

Figure 1.7 Derivation of the half-bridge converter: (a) two single-switch forward converters connected in series at the input side and in parallel at the secondary rectifiers, sharing the freewheeling diode and the output filter, (b) the two transformers sharing a magnetic core, (c) the positions of Q_1 and primary winding N_s interchanged, and (d) the final configuration of the half-bridge converter

Q_1 and Q_2 conduct in turn. Since the nonpolarity-marked terminals of the two primary windings are connected, the polarity-marked terminals have the same voltage potential and can be connected. It is obvious that the two primary windings are in parallel, and one of them can be removed. Figure 1.7d shows the final configuration of the converter, which is the half-bridge converter. As in the case of the push–pull converter, the freewheeling diode D_{FW} can be removed and D_{R1} and D_{R2} conduct simultaneously when Q_1 and Q_2 are off.

The half-bridge converter is equivalent to two single-switch forward converters connected in series at the input side. Thus, the voltage applied on the input side of each forward converter is half the input voltage $V_{in}/2$ and the magnitude of the primary winding voltage is $V_{in}/2$, which is half that of the push–pull converter. Key waveforms of the half-bridge converter are given in Figure 1.6, with V_{in} replaced by $V_{in}/2$. As in the push-pull converter, the transformer of the half-bridge converter is bidirectionally magnetized.

The voltage stress of the power switches in the half-bridge converter is $2 \cdot V_{in}/2 = V_{in}$; in fact, it can be seen from Figure 1.7d that when either switch is conducting, the other must withstand V_{in}.

1.2.4 Full-Bridge Converter

The dual-switch forward converter was derived in Section 1.2.1 and is redrawn in Figure 1.8a for convenience. This converter has an alternative configuration, shown in Figure 1.8b. When both switches conduct, the transformer is negatively magnetized; when the switches are off, the transformer is magnetically reset through diodes D_1 and D_4. The two kinds of dual-switch forward converter can be connected in parallel at the input sides and the output rectifier, as shown in Figure 1.8c, where $Q_1(Q_4)$ and $Q_2(Q_3)$ operate at the same frequency and with a time difference of $T_s/2$. If the two transformers share a magnetic core and an antiparallel diode is connected to each switch, as shown in Figure 1.8d, then the transformer can be magnetically reset through primary winding N_{p2} and the antiparallel diodes of Q_2 and Q_3, or through primary winding N_{p1} and the antiparallel diodes of Q_1 and Q_4. Consequently, diodes D_1 to D_4 can be removed. Since the two primary windings have the same voltage waveforms, points A_1 and A_2 and points B_1 and B_2 can be connected, respectively. Thus, the two primary windings are in parallel, and one can be removed, resulting in the circuit shown in Figure 1.8e. This is the full-bridge converter. Similarly, the freewheeling diode D_{FW} can be removed. For brevity of illustration, the antiparallel diodes of Q_1 to Q_4 are labeled D_1 to D_4.

The transformer of the full-bridge converter is bidirectionally magnetized, and the magnitude of the primary winding voltage is V_{in}, which is the same as that of the push–pull converter and twice that of the half-bridge converter. The voltage stress of the power switches of the full-bridge converter is V_{in}, which is the same as that of the dual-switch forward converter.

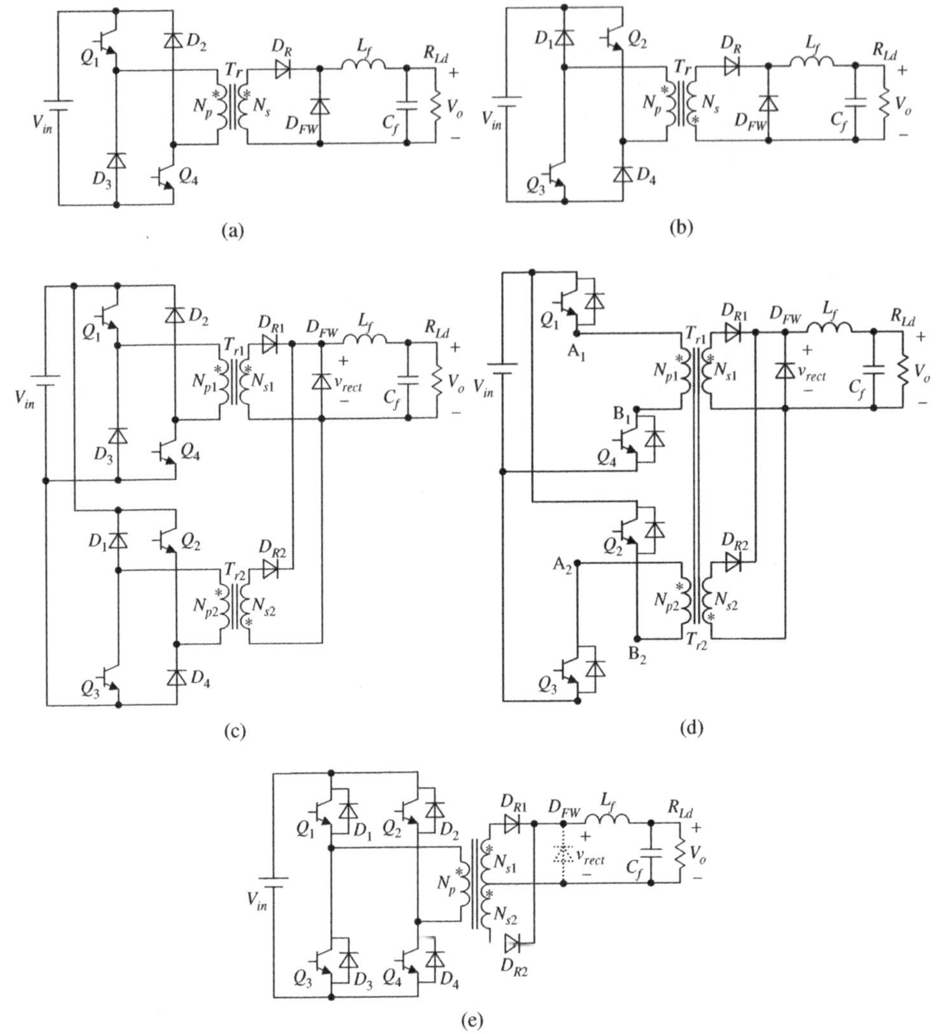

Figure 1.8 Derivation of the full-bridge converter: (a) dual-switch forward converter, (b) alternative configuration of the dual-switch forward converter, (c) two kinds of dual-switch forward converter connected in parallel at the input sides and the output rectifier, (d) the two transformers sharing a magnetic core, and (e) final configuration of the full-bridge converter

1.2.5 Comparison of Isolated Buck-Derived Converters

From the derivation of the forward converter (including single-switch and dual-switch versions), the push–pull converter, the half-bridge converter, and the full-bridge converter, it can be concluded that all of these isolated converters are originated from the buck converter. Table 1.1 provides a comparison.

Table 1.1 Comparison of isolated buck-derived converters

Converter type	Voltage stress of power switches	Primary-to-secondary-winding-turns ratio	Current stress of power switches	Number of power switches	Total power handling capacity of power switches	Ripple frequency of secondary rectified voltage	Maximum duty cycle of secondary rectified voltage
Single-switch forward	$2V_{in}$	K_0	I_o/K_0	1	$2V_{in}I_o/K_0$	f_s	0.5
Dual-switch forward	V_{in}	K_0	I_o/K_0	2	$2V_{in}I_o/K_0$	f_s	0.5
Push–pull	$2V_{in}$	$2K_0$	$I_o/(2K_0)$	2	$2V_{in}I_o/K_0$	$2f_s$	1
Half-bridge	V_{in}	K_0	I_o/K_0	2	$2V_{in}I_o/K_0$	$2f_s$	1
Full-bridge	V_{in}	$2K_0$	$I_o/(2K_0)$	4	$2V_{in}I_o/K_0$	$2f_s$	1

1. **Voltage stress of power switches:** The power switches of the single-switch forward converter and the push–pull converter have to withstand twice the input voltage, while the power switches of the dual-switch forward converter, the half-bridge converter, and the full-bridge converter are required to withstand the input voltage. Thus, the single-switch forward converter and the push–pull converter are suitable for low-voltage applications, while the dual-switch forward converter, the half-bridge converter, and the full-bridge converter are suitable for high-voltage applications.
2. **Transformer primary-to-secondary-winding-turns ratio:** The maximum duty cycle of the forward converter (including single-switch and dual-switch versions) is limited to 0.5 and the maximum duty cycle of the push–pull converter, the half-bridge converter, and the full-bridge converter can reach unity. Under conditions of the same input and output voltages, if the transformer winding turns ratio of the forward converter is K_0, the transformer-winding-turns ratios of the push–pull converter and of the full-bridge converter should be $2K_0$. For the half-bridge converter, although its duty cycle can reach unity, the magnitude of the primary voltage is half of the input voltage, so its transformer-winding-turns ratio is K_0.
3. **Current stress of power switches:** Neglecting the output filter inductor current ripple, the current stresses of the power switches of the forward converter and of the half-bridge converter are both I_o/K_0, while the current stresses of the power switches of the push–pull converter and of the full-bridge converter are both $I_o/(2K_0)$, where I_o is the output current.
4. **Total power handling capacity of power switches:** The power handling capacity of a power switch is defined as the product of the voltage stress and current stress imposed on the power switch. From the preceding analysis, it can readily be seen that the total power handling capacity (i.e., the number of power switches

multiplied by each power switch's power handling capacity) of the five isolated buck-derived converters is $2V_{in}I_o/K_0$. This means that for the same input and output, the total power handling capacities of switches of all five converters are the same. In other words, if the power switches have the same voltage and current ratings, the output power the converter can handle is proportional to the number of power switches. Of the five isolated buck-derived converters, the full-bridge converter has the most power switches (four) and the highest power handling capability. Therefore, the full-bridge converter has been widely used in medium-to-high-power-conversion applications.

5. **Output filter:** For the forward converter, the ripple frequency of the secondary rectified voltage is the switching frequency f_s and the maximum duty cycle is limited to 0.5. For the push–pull converter, the half-bridge converter, and the full-bridge converter, the secondary rectified voltage has a ripple frequency of $2f_s$ and a maximum duty cycle of 1. Therefore, for the same output voltage, the secondary rectified voltages of the push–pull converter, the half-bridge converter, and the full-bridge converter have smaller amounts of high-frequency harmonics than that of the forward converter, and thus the required output filter is much smaller.

1.3 Output Rectifier Circuits

Section 1.2 presented the derivations of the forward converter (including single-switch and dual-switch versions), the push–pull converter, the half-bridge converter, and the full-bridge converter. The output rectifier circuit of the forward converter is a half-wave rectifier circuit, while those of the push–pull converter, the half-bridge converter, and the full-bridge converter are full-wave rectifier circuits. In fact, since the push–pull converter, the half-bridge converter, and the full-bridge converter all transfer energy from the input to the load during both the positive and the negative half-periods, they can all also adopt the full-bridge rectifier circuit and the current-doubler rectifier circuit. In this section, the half-wave rectifier circuit, the full-bridge rectifier circuit, and the current-doubler rectifier circuit will be derived from the half-wave rectifier circuit. The purpose of this is to reveal the relationship among these output rectifier circuits.

1.3.1 Half-Wave Rectifier Circuit

Figure 1.9a and Figure 1.9b show the positive and negative half-wave rectifier circuits, respectively. The two half-wave rectifier circuits can only transfer energy to load in the positive or negative half-period of transformer primary voltage v_p. The key waveforms are depicted in Figure 1.10, from which it can be seen that:

1. The output voltage V_o can be derived as:

$$V_o = D_h V_{pm}/K \tag{1.4}$$

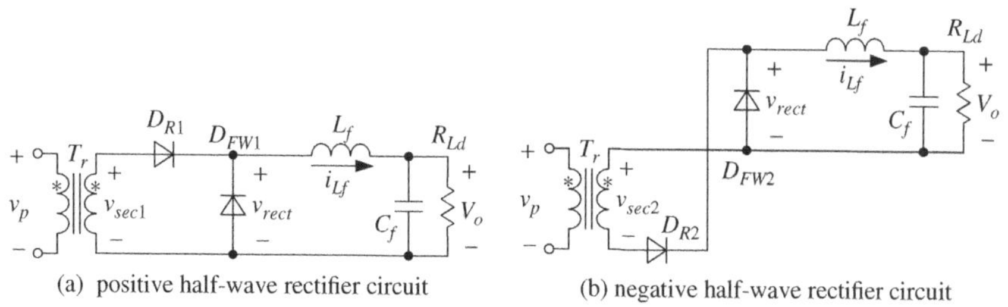

Figure 1.9 Two kinds of half-wave rectifier circuit

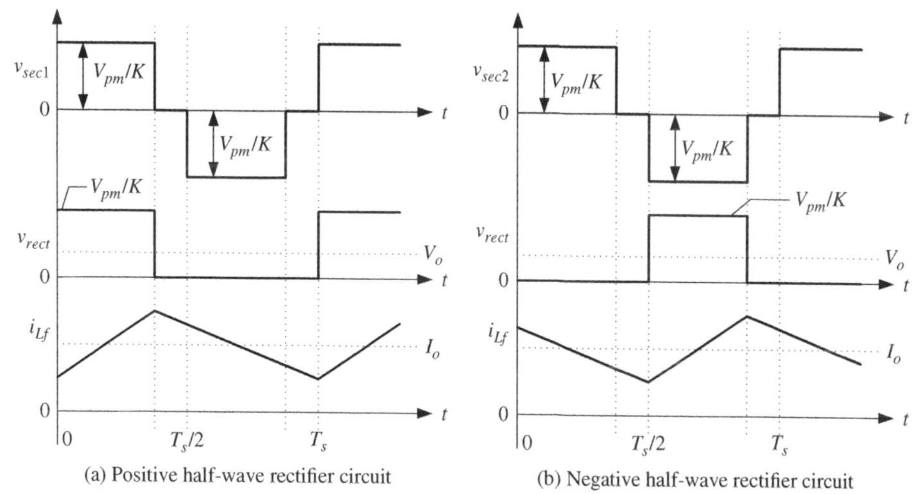

Figure 1.10 Waveforms of two kinds of half-wave rectifier circuit

where V_{pm} is the magnitude of the transformer primary voltage, D_h is the duty cycle of the half-wave rectifier circuit, which is the ratio of the width of the positive (or negative) half-period to the switching period, and K is the primary-to-secondary-winding-turns ratio.

2. The ripple frequency of the rectified voltage and output filter inductor current is the switching frequency.
3. The voltage stress of both the rectifier diode and the freewheeling diode is V_{pm}/K.

1.3.2 Full-Wave Rectifier Circuit

If the transformer is expected to transfer energy to the load during both positive and negative half-periods, the positive and negative half-wave rectifier circuits should be combined, as shown in Figure 1.11a. As illustrated in Section 1.2, when v_p is equal to zero, the output filter inductor can freewheel through freewheeling diodes D_{FW1} and D_{FW2} or through two secondary windings via D_{R1} and D_{R2}. Therefore, D_{FW1} and D_{FW2}

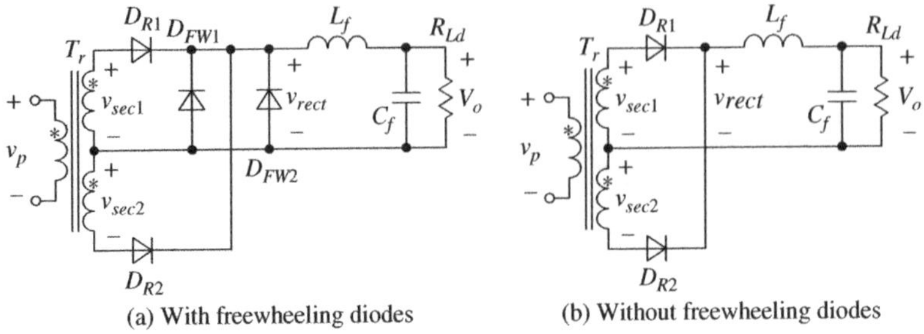

Figure 1.11 Full-wave rectifier circuit

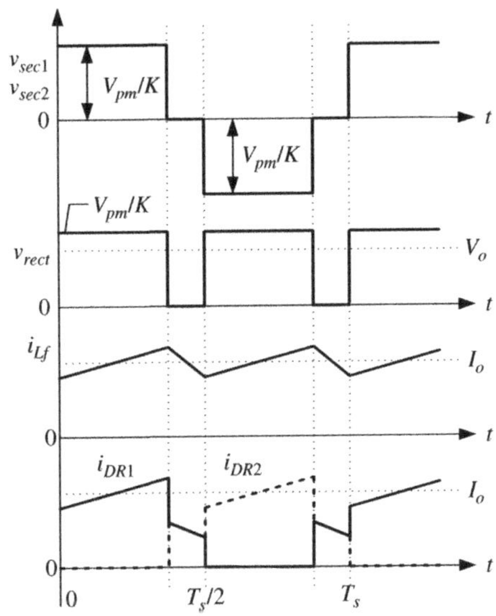

Figure 1.12 Key waveforms of the full-wave rectifier circuit

are redundant and can be removed, as shown in Figure 1.11b. Thus, the full-wave rectifier circuit is obtained. Figure 1.12 shows the key waveforms of the full-wave rectifier circuit, from which we see that:

1. The output voltage V_o is given by:

$$V_o = 2D_h V_{pm}/K = D_y V_{pm}/K \tag{1.5}$$

   where D_y is the duty cycle of the secondary rectified voltage, which is the ratio of the pulse width of the secondary rectified voltage to half the switching period. Here, D_y is twice the duty cycle of the half-wave rectifier circuit; that is, $D_y = 2D_h$.
2. The ripple frequency of the rectified voltage and the output filter inductor current is twice the switching frequency.

3. The voltage stress of the two rectifier diode is $2V_{pm}/K$. During the freewheeling period, the two rectifier diodes share the output filter inductor current.

From Equations 1.4 and 1.5, it can be seen that the output voltage of the full-wave rectifier circuit is twice that of the half-wave rectifier circuit when D_h is the same. This is because with the full-wave rectifier circuit, voltage is applied on the output filter in both the positive and the negative half-periods. If the output voltage stays the same, the transformer-turns ratio of the full-wave rectifier circuit is twice that of the half-wave rectifier circuit and the voltage stresses of the rectifier diodes of the two kinds of rectifier circuit are equal. Compared with the half-wave rectifier circuit, the ripple frequency of the secondary rectified voltage is doubled and the switching-frequency harmonics are significantly reduced. Thus, the output filter can be drastically reduced.

1.3.3 Full-Bridge Rectifier Circuit

In the full-wave rectifier circuit, the transformer has two secondary windings, with each one only conducting for a half-period. If the secondary winding can conduct current during both positive and negative half-periods, the utilization is increased and one secondary winding can be removed, leading to a simple configuration of the transformer. Figure 1.13a shows the rectifier circuit with only one secondary winding

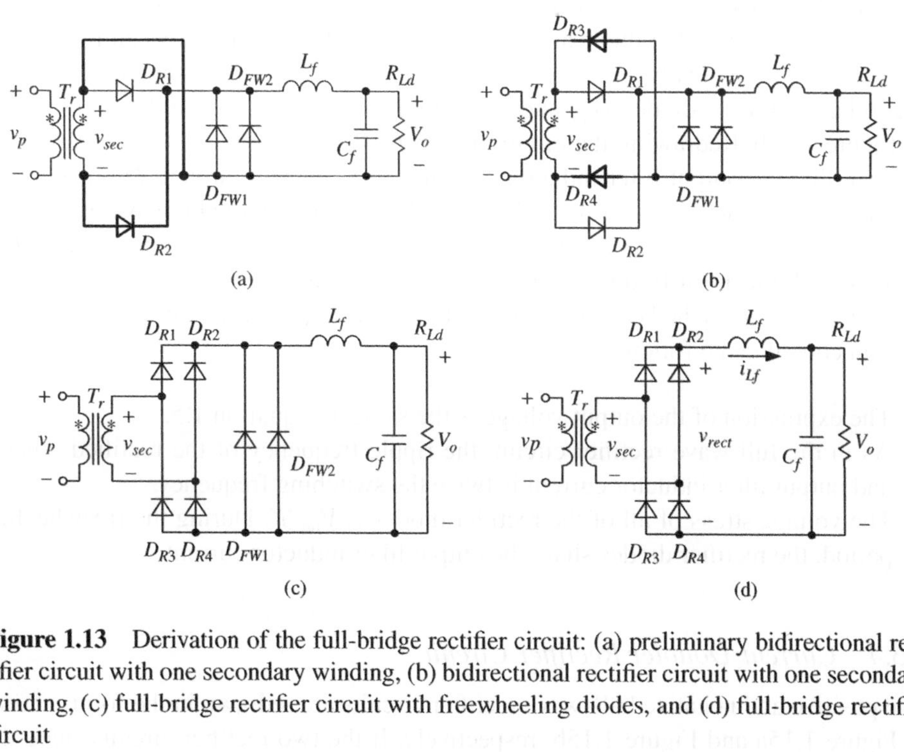

Figure 1.13 Derivation of the full-bridge rectifier circuit: (a) preliminary bidirectional rectifier circuit with one secondary winding, (b) bidirectional rectifier circuit with one secondary winding, (c) full-bridge rectifier circuit with freewheeling diodes, and (d) full-bridge rectifier circuit

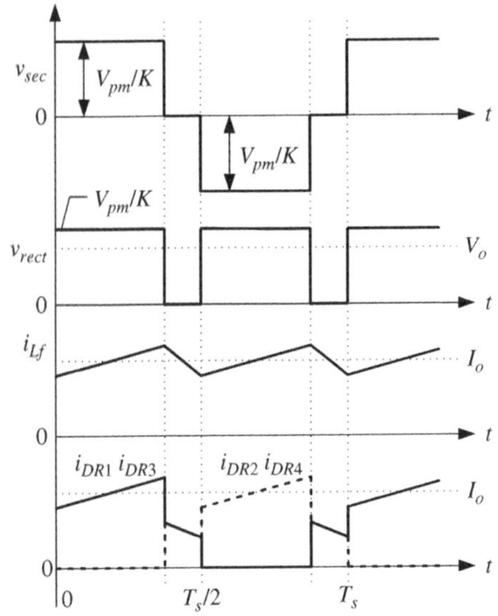

Figure 1.14 Key waveforms of the full-bridge rectifier circuit

conducting current bidirectionally, where D_{R1} and D_{FW1} are the rectifier diode and freewheeling diode of the positive half-wave rectifier circuit, respectively, and D_{R2} and D_{FW2} are the rectifier diode and freewheeling diode of the negative half-wave rectifier circuit, respectively. However, such an arrangement leads to short-circuit of the secondary winding; diodes D_{R4} and D_{R3} should be inserted to avoid this, as shown in Figure 1.13b. Redrawing the circuit in Figure 1.13b as Figure 1.13c, it can easily be seen that the output filter inductor current can freewheel through D_{FW1} or D_{FW2}, or through the branch consisting of D_{R1} and D_{R3}, or through the branch consisting of D_{R2} and D_{R4}. Therefore, D_{FW1} and D_{FW2} are redundant and can be removed, leading to the well-known full-bridge rectifier circuit, as shown in Figure 1.13d.

Figure 1.14 shows the key waveforms of the full-bridge rectifier circuit, from which we can conclude the following:

1. The expression of the output voltage is the same as Equation 1.5.
2. As in the full-wave rectifier circuit, the ripple frequency of the rectified voltage and output filter inductor current is twice the switching frequency.
3. The voltage stress of all of the rectifier diodes is V_{pm}/K. During the freewheeling period, the rectifier diodes share the output filter inductor current.

1.3.4 Current-Doubler Rectifier Circuit

The positive and negative half-wave rectifier circuits are redrawn in the forms shown in Figure 1.15a and Figure 1.15b, respectively. If the two rectifier circuits share the

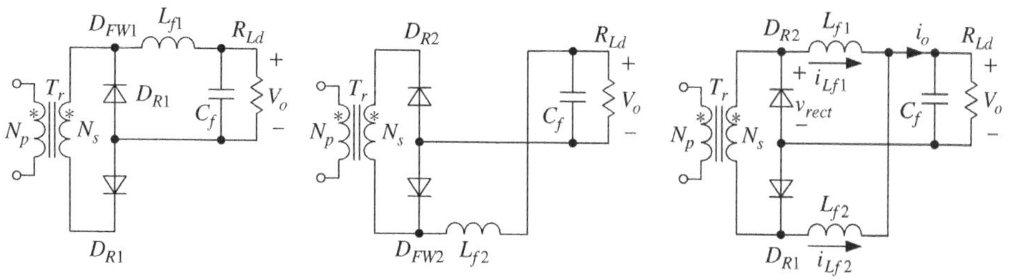

Figure 1.15 Derivation of the current-doubler rectifier circuit: (a) positive half-wave rectifier circuit, (b) negative half-wave rectifier circuit, and (c) current-doubler rectifier circuit

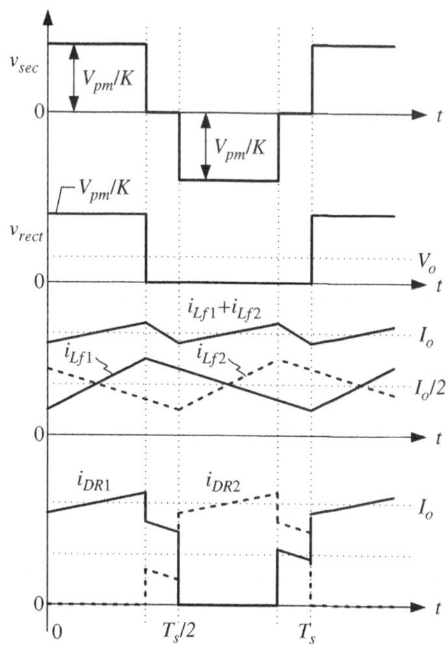

Figure 1.16 Key waveforms of the current-doubler rectifier circuit

same set of transformer, rectifier diode, and freewheeling diode, the current-doubler rectifier circuit is obtained, as shown in Figure 1.15c, where two output filter inductor currents are supplied to the output filter capacitor and load. Figure 1.16 shows the key waveforms of the current-doubler rectifier circuit, whose characteristics are as follows:

1. It can be treated as the parallel of the positive and negative half-wave rectifier circuits. The expression of the output voltage is thus the same as that of the

half-wave rectifier circuit; that is, Equation 1.4. In other words, with the same transformer-turns ratio, the output voltage of the current-doubler rectifier circuit is half that of the full-wave rectifier circuit or the full-bridge rectifier circuit.
2. The two output filter inductor currents pulsate at the switching frequency with a phase shift of 180° and the output current i_o is the sum of the two output filter inductor currents and ripples at twice the switching frequency. Since two output filter inductor currents have a phase shift of 180°, the current ripples at the switching frequency and its odd multiples are cancelled in the output current, which means that the ripple of i_o is smaller than the individual output filter inductor current ripple.
3. The voltage stress of the two rectifier diodes is V_{pm}/K. Under conditions of the same input and output voltages, and with the same duty cycle D_h, the transformer-winding-turns ratio of the current-doubler rectifier circuit is half that of the full-wave rectifier circuit and the full-bridge rectifier circuit. Therefore, the voltage stress of the rectifier diodes is equal to that of the full-wave rectifier circuit and double that of the full-bridge rectifier circuit.

From this analysis, it can be concluded that the full-wave rectifier circuit, the full-bridge rectifier circuit, and the current-doubler rectifier circuit can be derived from the half-wave rectifier circuit. Table 1.2 compares the three kinds of rectifier circuit, which can be described as follows:

1. **Transformer primary-to-secondary-winding-turns ratio:** The full-wave rectifier circuit and the full-bridge rectifier circuit can be treated as a series connection of positive and negative half-wave rectifier circuits, while the current-doubler rectifier circuit can be treated as a parallel connection of positive and negative half-wave rectifier circuits. Therefore, when operating with the same duty cycle D_h and transformer primary-to-secondary-winding-turns ratio, the output voltage of the full-wave rectifier circuit and the full-bridge rectifier circuit is twice that of the current-doubler rectifier circuit. In other words, if under the same input and output voltages, the transformer-turns ratio of the full-wave rectifier circuit and the full-bridge rectifier circuit is K_0, that of the current-doubler rectifier circuit should be $K_0/2$.

Table 1.2 Comparison of three kinds of rectifier circuit

Rectifier circuit type	Transformer primary-to-secondary-turns ratio	Voltage stress of rectifier diodes	Current stress of rectifier diodes	Number of rectifier diodes	Total power handling capacity of rectifier diodes
Full-wave rectifier	K_0	$2V_{in}/K_0$	I_o	2	$4V_{in}I_o/K_0$
Full-bridge rectifier	K_0	V_{in}/K_0	I_o	4	$4V_{in}I_o/K_0$
Current-doubler rectifier	$K_0/2$	$2V_{in}/K_0$	I_o	2	$4V_{in}I_o/K_0$

2. **Voltage stress of rectifier diodes:** The voltage stress of the rectifier diodes of the full-wave rectifier circuit and the current-doubler rectifier circuit is $2V_{in}/K_0$, while that of the rectifier diodes of the full-bridge rectifier circuit is V_{in}/K_0, where V_{in} is the input voltage.
3. **Current stress of rectifier diodes:** Neglecting the output filter inductor current ripple, the current stress of the rectifier diodes in all three rectifier circuits is the output current I_o.
4. **Total power handling capacity of rectifier diodes:** The power handling capacity of a rectifier diode is defined as the product of the voltage stress and the current stress of the rectifier diode. From the preceding discussion, it can be seen that the total power handling capacity of the rectifier diodes (i.e., the power handling capacity of each rectifier diode multiplied by the number of rectifier diodes in all three rectifier circuits) is equal to $4V_{in}I_o/K_0$.

1.4 Basic Operating Principle of Full-Bridge Converters

1.4.1 Topologies of Full-Bridge Converters

Since the transformer of the full-bridge converter transfers energy from the input to the load during both the half-periods, the rectifier circuit of the full-bridge converter can be either the full-wave rectifier circuit, the full-bridge rectifier circuit, or the current-doubler rectifier, as shown in Figure 1.17.

1.4.2 Pulse-Width Modulation Strategies for Full-Bridge Converters

Figure 1.18 shows the commonly used PWM strategies for the full-bridge converter. In the basic PWM strategy, the two diagonal switches of the two legs turn on and off simultaneously, as shown in Figure 1.18a. The two switches in one bridge leg can also be operated in a PWM fashion, and the two in the other bridge leg in a complementary manner with 50% duty cycle, as shown in Figure 1.18b. Figure 1.18c shows the well-known phase-shift control, where the switches of each bridge leg operate complementarily with 50% duty cycle and a phase-shift is introduced between the two legs. Regulation of the output voltage can thus be achieved by controlling the phase-shift.

1.4.3 Basic Operating Principle of a Full-Bridge Converter with a Full-Wave Rectifier Circuit and a Full-Bridge Rectifier Circuit

Regardless of the kind of PWM strategy adopted, the operating principle of the full-bridge converter with a full-wave rectifier circuit is the same. In the following, the basic strategy is used for illustration. Figure 1.19 shows the key waveforms of the

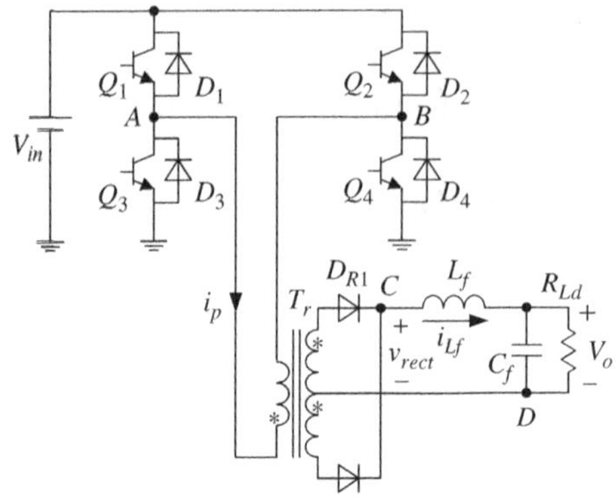

(a) With full-wave rectifier circuit

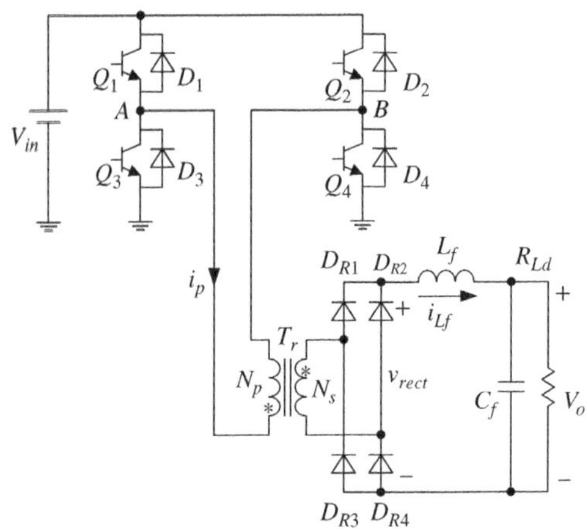

(b) With full-bridge rectifier circuit

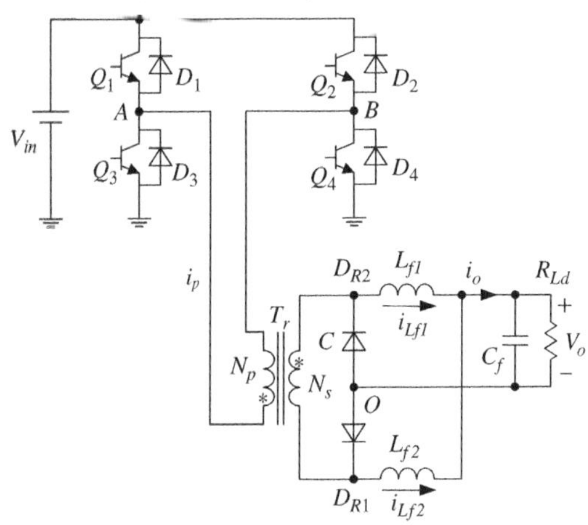

(c) With current-doubler rectifier circuit

Figure 1.17 Full-bridge converter with different output rectifier circuits

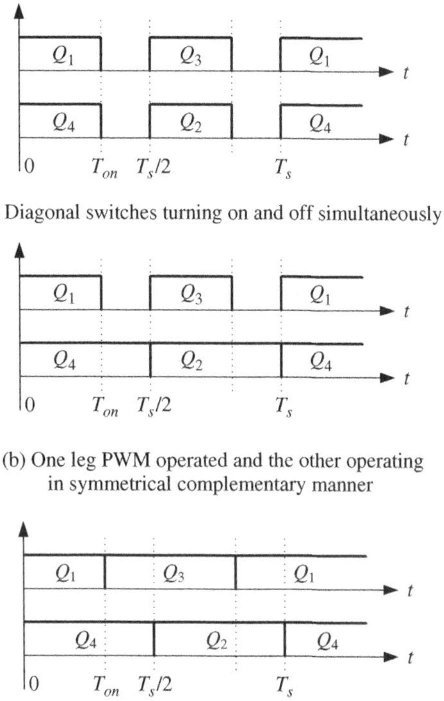

(a) Diagonal switches turning on and off simultaneously

(b) One leg PWM operated and the other operating in symmetrical complementary manner

(c) Phase-shifted control

Figure 1.18 Widely used modulation strategies for the full-bridge converter

full-bridge converter with a full-wave rectifier circuit, while the equivalent circuits of the topological modes are depicted in Figure 1.20.

When the diagonal switches Q_1 and Q_4 are conducting, as shown in Figure 1.20a, the voltage across the midpoints of the bridge legs v_{AB} is equal to V_{in}, the secondary rectifier diode D_{R1} conducts, and the secondary rectified voltage v_{rect} is equal to V_{in}/K, where K is the transformer primary-to-secondary-winding-turns ratio. The voltage applied across the output filter inductor L_f is $V_{in}/K - V_o$, which causes the inductor current i_{Lf} to increase linearly. The primary current i_p is equal to the output filter inductor current reflected to the primary side (i.e., $i_p = i_{Lf}/K$), and correspondingly it increases linearly. Also, i_p flows through Q_1 and Q_4.

When the diagonal switches Q_2 and Q_3 are conducting, as shown in Figure 1.20b, $v_{AB} = -V_{in}$, D_{R2} conducts, $v_{rect} = V_{in}/K$, and i_{Lf} increases linearly.

When all of the power switches are off, the primary current i_p is zero and the output filter inductor current freewheels through the two rectifier diodes as shown in Figure 1.20c. The two rectifier diodes shares the output filter inductor current (i.e., $i_{DR1} = i_{DR2} = i_{Lf}/2$). Since both rectifier diodes are conducting, the two secondary winding voltages are clamped to zero. Thus, $v_{rect} = 0$ and the voltage applied to the output filter inductor is $-V_o$. This negative voltage makes i_{Lf} decay linearly. If the

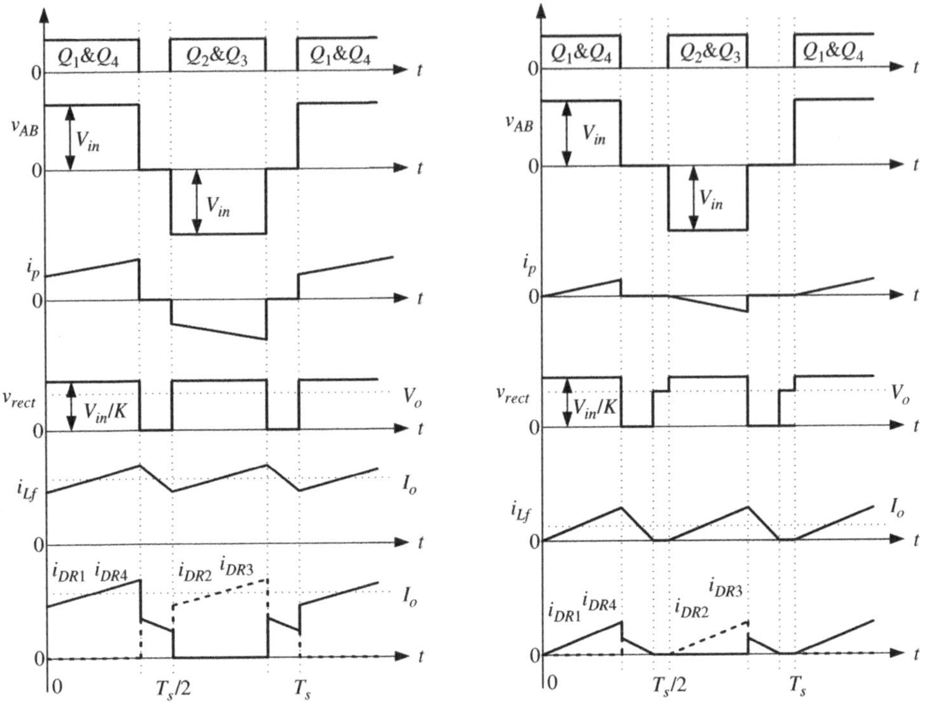

(a) Continuous current mode (b) Discontinuous current mode

Figure 1.19 Key waveforms of the full-bridge converter with full-wave rectifier circuit

load is light or the output filter inductor is small, i_{Lf} may decay to zero before one pair of diagonal switches is turned on; it is then kept at zero, as shown in Figure 1.20d. Under this condition, the full-bridge converter operates in discontinuous current mode (DCM). The corresponding waveforms are sketched in Figure 1.19b.

The operating principle of the full-bridge converter with a full-bridge rectifier circuit is similar to that of one with a full-wave rectifier circuit. The key waveforms are given in Figure 1.19, where the rectifier diode current waveforms are shown. For the equivalent circuits, the primary side is the same, while the secondary side is shown in the dashed block in Figure 1.20. The difference between the two rectifier circuits is that, for the full-bridge rectifier circuit, when all of the power switches are off, the output filter inductor current freewheels through the four rectifier diodes, while the secondary winding current is zero.

1.4.4 Basic Operating Principle of a Full-Bridge Converter with a Current-Doubler Rectifier Circuit

The full-bridge converter with a current-doubler rectifier circuit is shown in Figure 1.17c. The operation of a converter adopting the basic PWM strategy (see

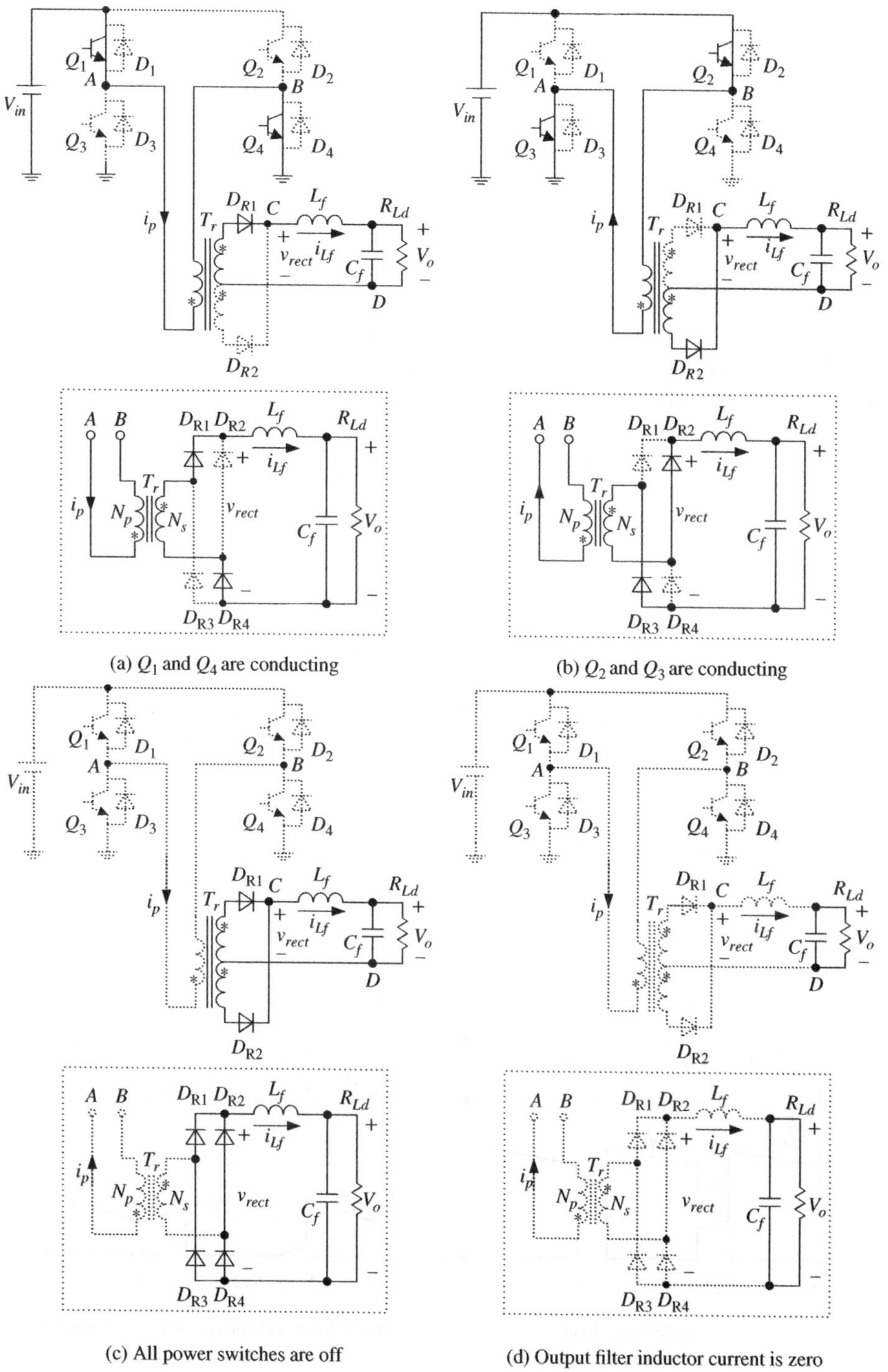

(a) Q_1 and Q_4 are conducting

(b) Q_2 and Q_3 are conducting

(c) All power switches are off

(d) Output filter inductor current is zero

Figure 1.20 Equivalent circuit of switching modes of the full-bridge converter

Figure 1.18a) is different from that of one adopting either of the other two PWM strategies (see Figure 1.18b,c).

1.4.4.1 Basic PWM Strategy

At Heavy Load

The key waveforms of the full-bridge converter with a current-doubler rectifier circuit under heavy load condition are shown in Figure 1.21a. When the diagonal power switches Q_1 and Q_4 conduct, the secondary rectifier diode D_{R1} also conducts, as shown in Figure 1.22a. During this interval, $v_{AB} = V_{in}$ and $v_{CO} = V_{in}/K$. The voltage across L_{f1} is $V_{in}/K - V_o$ and i_{Lf1} increases linearly, while the voltage across L_{f2} is $-V_o$ and i_{Lf2} decreases linearly. The primary current i_p is equal to i_{Lf1} reflected to the primary side (i.e., $i_p = i_{Lf1}/K$) and increases linearly. The output current i_o is the sum of the two output filter inductor currents (i.e., $i_o = i_{Lf1} + i_{Lf2}$) and also increases linearly.

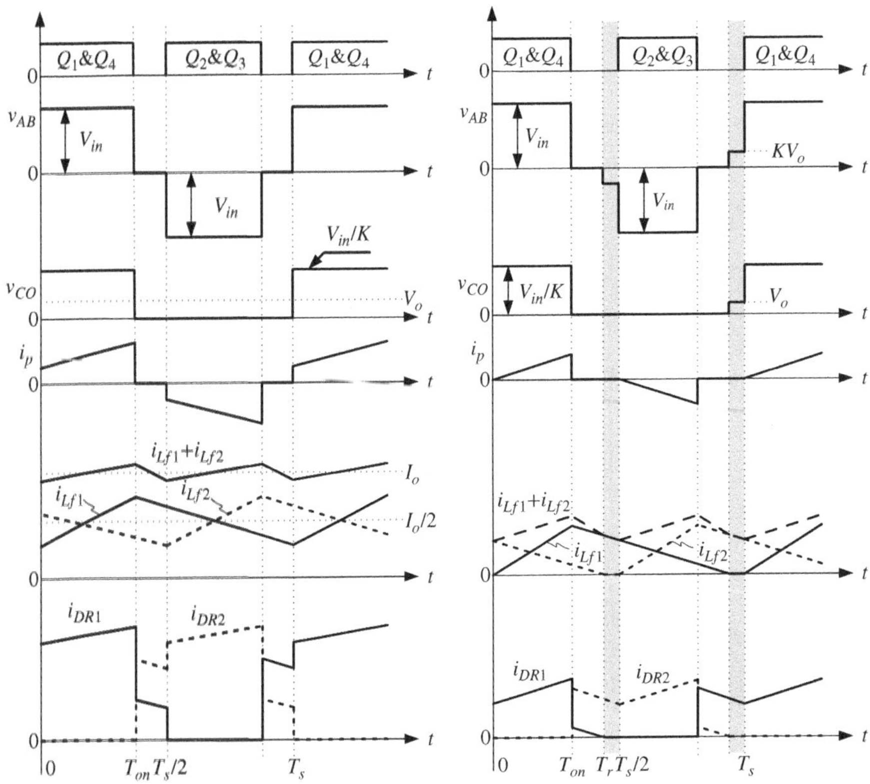

(a) At heavy load

(b) At light load and i_{Lf1} and i_{Lf2} keep positive

Figure 1.21 Key waveforms of the full-bridge converter with current-doubler rectifier circuit under the basic PWM strategy

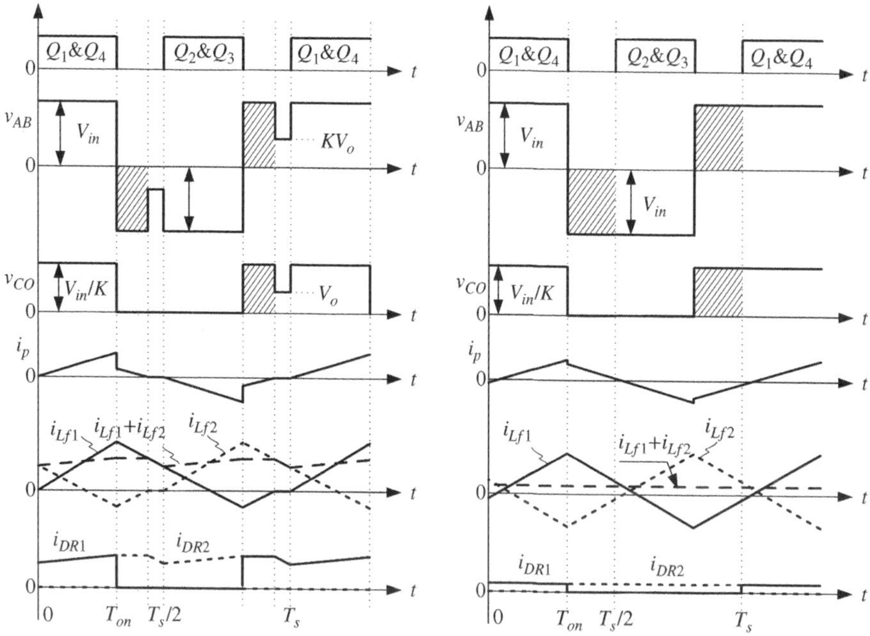

(c) i_{Lf1} and i_{Lf2} are bidirectional

(d) Discontinuous current mode

Figure 1.21 (*Continued*)

When Q_1 and Q_4 are turned off, i_{Lf1} and i_{Lf2} flow through D_{R2} and D_{R1}, respectively, and i_p becomes zero, as shown in Figure 1.22b. At this point, the voltages across the two output filter inductors are all $-V_o$, making i_{Lf1} and i_{Lf2} decay linearly. Since both D_{R1} and D_{R2} are conducting, $v_{CO}=0$ and the secondary winding voltage is zero. Correspondingly, the primary voltage $v_{AB}=0$.

The operation when Q_2 and Q_3 are conducting is similar to that when Q_1 and Q_4 are conducting, which is omitted here.

At Light Load when Output Filter Inductor Currents are Positive as Diagonal Switches Conduct

As just discussed, when Q_1 and Q_4 are turned off, i_{Lf1} and i_{Lf2} freewheel through D_{R2} and D_{R1}, respectively, and both decay linearly, as shown in Figure 1.22b. If the load is very light or the output filter inductor is not large enough, i_{Lf2} will decay to zero and D_{R1} will turn off naturally, as shown in Figure 1.22c. Following this, i_{Lf2} is kept at zero until Q_2 and Q_3 are turned on, as shown in the dashed lines within interval $[T_r, T_s/2]$ in Figure 1.21b. During this interval, $v_{AB}=-KV_o$ and $v_{CO}=0$. Similarly, when i_{Lf1} decays to zero, it stays at zero, as shown in the dashed lines in Figure 1.21b. During this interval, $v_{AB}=KV_o$ and $v_{CO}=V_o$.

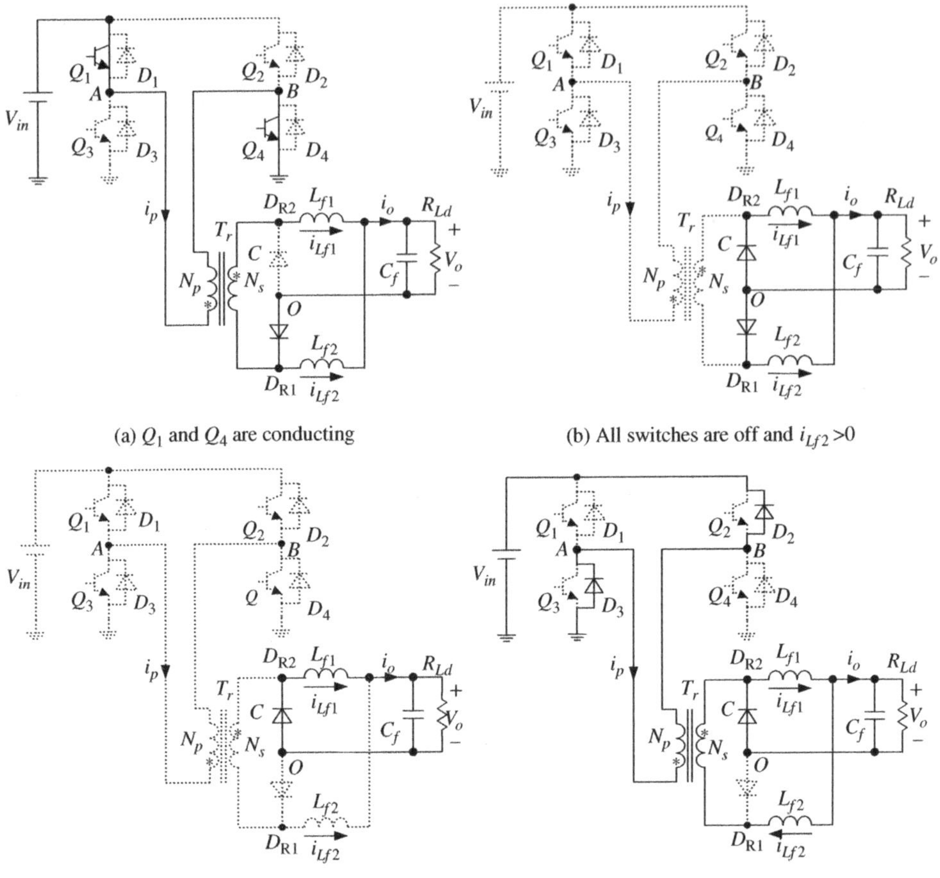

(a) Q_1 and Q_4 are conducting

(b) All switches are off and $i_{Lf2} > 0$

(c) All switches are off and $i_{Lf2} = 0$

(d) All switches are off and $i_{Lf2} < 0$

Figure 1.22 Equivalent circuits of switching modes of the full-bridge converter with current-doubler rectifier circuit under the basic PWM strategy

At Light Load when Output Filter Inductor Current becomes Negative as Diagonal Switches Conduct

When Q_1 and Q_4 are conducting, i_{Lf1} increases, i_{Lf2} decays, and i_p is equal to the reflected i_{Lf1} (i.e., $i_p = i_{Lf1}/K$). If the load becomes lighter, i_{Lf2} will cross zero and continue flowing in the negative direction. When Q_1 and Q_4 are turned off, i_{Lf1} freewheels through D_{R2} and decays linearly. Moreover, i_{Lf2} flows through the secondary winding and is reflected to the primary side (i.e., $i_p = -i_{Lf2}/K$). Thus, i_p is positive and flows through D_2 and D_3, as shown in Figure 1.22d. During this interval, $v_{AB} = -V_{in}$ and the voltage across L_{f2} is $V_{in}/K - V_o$, forcing i_{Lf2} to increase linearly. When i_{Lf2} increases to zero, i_p reduces to zero correspondingly. Since all the power switches are off, i_p cannot flow in the reverse direction and must be kept at zero, which makes i_{Lf2} stay at zero until Q_2 and Q_3 are turned on. It can be seen from Figure 1.21c that, compared with Figure 1.21b, there are additional portions in v_{AB} and v_{CO}, shown as the dashed area. Thus, V_o is equal to the average value of v_{CO}, and clearly V_o is dependent not only on the duty cycle but also on the load.

If the load continues to reduce, the hatched area shown in Figure 1.21c increases to reach the turn-on instant of diagonal two switches, as shown in Figure 1.21d. This means that at the turn-on instant of Q_2 and Q_3, i_{Lf2} is still negative. When i_{Lf2} increases to zero, it can continue to flow in the positive direction, and it is continuous. Under this condition, v_{AB} becomes an ac square voltage with 180° electrical degree, while v_{CO} becomes a pulse voltage with a magnitude of V_{in}/K and a pulse width of $T_s/2$. The average value of v_{CO} is equal to $V_{in}/(2K)$. Since V_o is the average value of v_{CO}, it is equal to $V_{in}/(2K)$ and is independent of the duty cycle.

This analysis illustrates that, employing the basic PWM strategy, the output voltage of the full-bridge converter with a current-doubler rectifier circuit is dependant not only on the duty cycle but also on the load. Furthermore, the output voltage will lose control at light load. Therefore, the basic PWM strategy is not suitable for the full-bridge converter with a current-doubler rectifier circuit.

1.4.4.2 Phase-Shifted Control

Figure 1.23 shows the key waveforms of the full-bridge converter with a current-doubler rectifier circuit when the phase-shifted control is adopted. At heavy load, the operation is the same as that adopting the basic PWM strategies, as shown in Figure 1.23a.

At light load, when Q_1 and Q_4 are conducting, i_{Lf2} decays and becomes negative. When the leading-leg switches Q_1 is turned off and Q_3 is turned on, i_{Lf1} is freewheeling through D_{R2} and decays linearly and i_{Lf2} flows through the secondary winding and is reflected to the primary side (i.e., $i_p = -i_{Lf2}/K$). Here, i_p flows through D_3 and Q_4, as shown in Figure 1.24a. At this time, $v_{AB} = 0$ and the voltage across L_{f2} is still $-V_o$, meaning i_{Lf2} continues to decay. Figure 1.23b shows the key waveforms under this condition.

If the load becomes lighter, the two output filter inductor currents decay linearly when $v_{AB} = 0$. When $i_{Lf1} = -i_{Lf2}$, D_{R2} is turned off, and i_{Lf1} and i_{Lf2} are kept unchanged. Thus, $i_o = 0$. At this time, the full-bridge converter operates in DCM. It should be noted that, for the current-doubler rectifier circuit, "DCM" refers to the sum of the two output filter inductor currents flowing into the load being in discontinuous conduction; the two output filter inductor currents are not zero. Correspondingly, continuing current mode (CCM) refers to when the sum of two output filter inductor currents is continuous.

From this analysis, the following conclusions can be drawn:

1. When the phase-shifted control is adopted, regardless of whether the operation is in CCM or DCM, the output voltage of the full-bridge converter with a

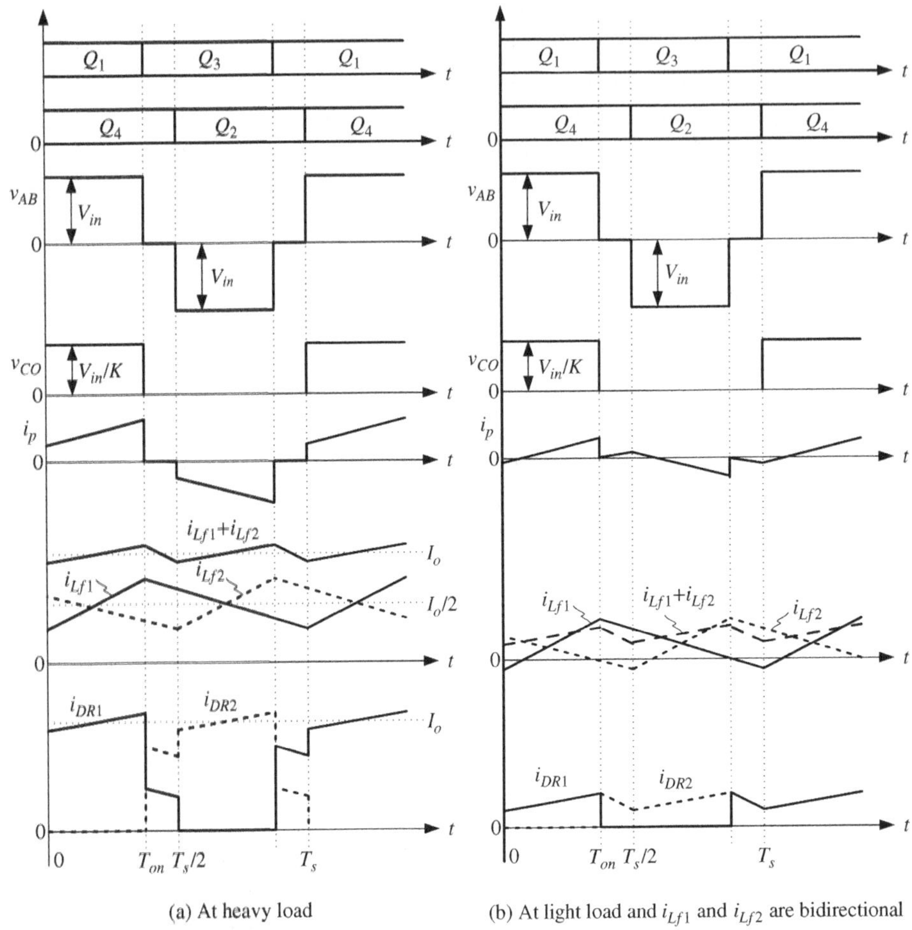

Figure 1.23 Key waveforms of the full-bridge converter with current-doubler rectifier circuit under phase-shifted control

current-doubler rectifier circuit can be regulated by controlling the duty cycle. This is different from the basic PWM strategy.

2. If the output filter inductor current becomes negative in active mode ($v_{AB}=+V_{in}$ or $-V_{in}$), it will be reflected to the primary side when $v_{AB}=0$. The induced primary current increases linearly. This can be used to achieve ZVS for the lagging leg, which will be discussed in Chapter 8.

The operation of the full-bridge converter with a current-doubler rectifier circuit employing the PWM modulation strategy shown in Figure 1.18b is the same as that using a phase-shifted control. Details are omitted here.

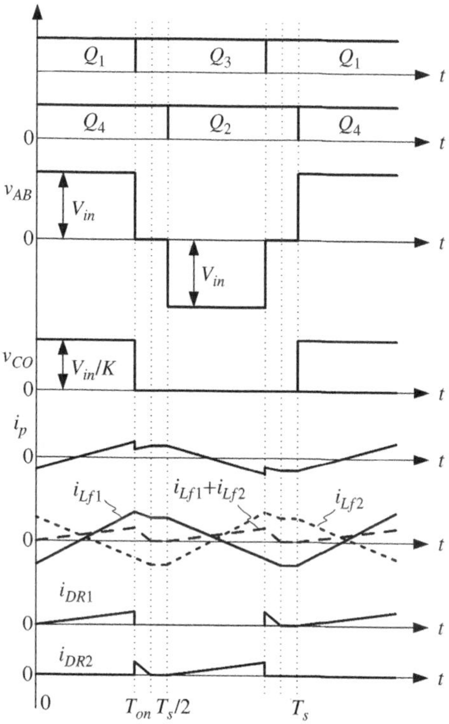

(c) DCM

Figure 1.23 (*Continued*)

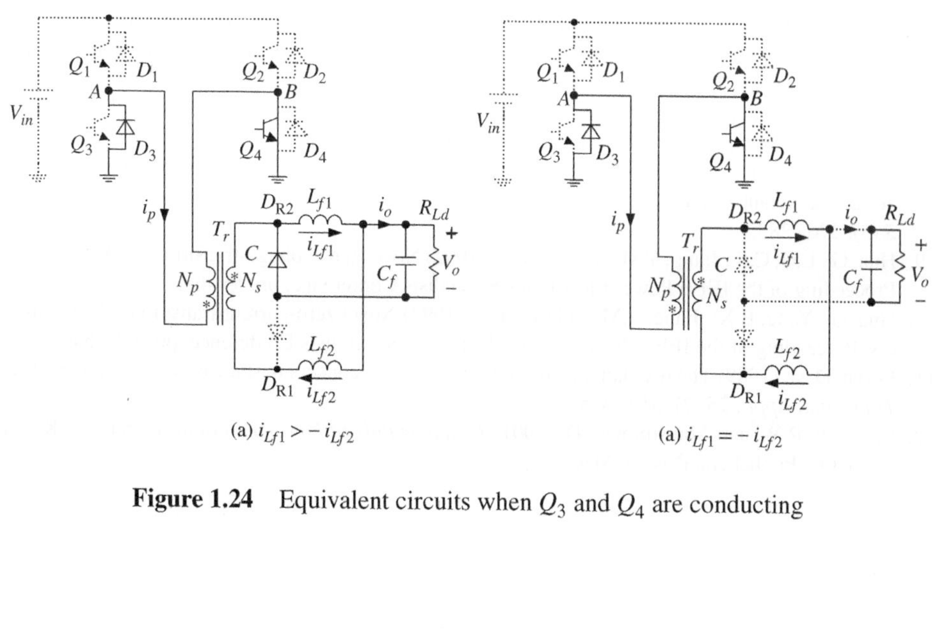

(a) $i_{Lf1} > -i_{Lf2}$ (a) $i_{Lf1} = -i_{Lf2}$

Figure 1.24 Equivalent circuits when Q_3 and Q_4 are conducting

1.5 Summary

This chapter introduced some development trends in the switching techniques, classifications, and requirements of power electronics converters, as well as the types and characteristics of dc–dc converters. The forward converter (including single-switch and dual-switch versions), push–pull converter, half-bridge converter, and full-bridge converter were derived from the buck converter, in order to help readers understand more clearly the relationships among various isolated buck-derived converters. Meanwhile, the full-wave rectifier circuit, the full-bridge rectifier circuit, and the current-doubler rectifier circuit were derived from the half-wave rectifier circuit. Again, the emphasis was on the relationships among these rectifier circuits. The basic operation of the full-bridge converter with full-wave rectifier circuit, full-bridge rectifier circuit, and current-doubler rectifier circuit was analyzed. The basic background necessary for study of the operation and design of soft-switching PWM full-bridge converters has thus been provided.

References

1. Spiazzi, G., Buso, S., Citron, M. *et al.* (2003) Performance evaluation of a Schottky SiC power diode in a boost PFC application. *IEEE Trans. Power Electron.*, **18** (6), 1249–1253.
2. Zhang, Q., Callanan, R., Das, M.K. *et al.* (2010) SiC power devices for microgrids. *IEEE Trans. Power Electron.*, **25** (12), 2889–2896.
3. Oruganti, R. and Lee, F.C. (1985) Resonant power processors, part I: state plane analysis. *IEEE Trans. Ind. Appl.*, **21** (6), 1453–1461.
4. Oruganti, R. and Lee, F.C. (1985) State-plane analysis of parallel resonant converters. Proceeding of the IEEE Power Electronics Specialists Conference, pp. 56–73.
5. Lee, F.C. (1988) High-frequency quasi-resonant and multi-resonant converter technologies. Proceeding of the IEEE International Conference on Industrial Electronics, pp. 509–521.
6. Tabisz, W.A. and Lee, F.C. (1988) Zero-voltage-switching multi-resonant technique – A novel approach to improve performance of high frequency quasi-resonant converters. Proceeding of the IEEE Power Electronics Specialists Conference, pp. 9–17.
7. Hua, G. and Lee, F.C. (1991) A new class of zero-voltage-switched PWM converters. Proceeding of the High Frequency Power Conversion Conference, pp. 244–251.
8. Barbi, I., Bolacell, J.C., Martins, D.C. and Libano, F.B. (1990) Buck quasi-resonant converter operating at constant frequency: analysis, design and experimentation. *IEEE Trans. Power Electron.*, **5** (3), 276–283.
9. Hua, G., Leu, C.S., Jiang, Y.M. and Lee, F.C. (1992) Novel zero-voltage-transition PWM converters. Proceeding of the IEEE Power Electronics Specialists Conference, pp. 55–61.
10. Hua, G., Yang, E.X., Jiang, Y.M. and Lee, F.C. (1993) Novel zero-current-transition PWM converters. Proceeding of the IEEE Power Power Electronics Specialists Conference, pp. 538–544.
11. Divan, D.M. (1989) The resonant dc link converter – a new concept in static power conversion. *IEEE Trans. Ind. Appl.*, **25** (2), 317–325.
12. Erickson, R.W. and Maksimovic, D. (2001) *Fundamentals of Power Electronics*, 2nd edn, Kluwer Academic Publishers, Boston, MA.

2

Theoretical Basis of Soft Switching for PWM Full-Bridge Converters

In Chapter 1, the basic circuit configurations and control of full-bridge converters were introduced. In this chapter and subsequent ones, the pulse-width modulation (PWM) soft-switching techniques for full-bridge converters will be expounded. The soft-switching techniques to be discussed are applicable to converters that are controlled under a PWM strategy. This chapter reviews the theoretical basis of the various PWM soft-switching techniques applied to full-bridge converters [1].

2.1 PWM Strategies for Full-Bridge Converters

2.1.1 Basic PWM Strategy

The basic circuit configuration of the full-bridge converter and its key waveforms are shown in Figure 2.1, where V_{in} is the input dc voltage source, T_r is a high-frequency isolation transformer with primary-and-secondary-windings ratio K, D_{R1} and D_{R2} constitute the output full-wave rectifier, L_f is the output filter inductor, C_f is the output capacitor, and R_{Ld} is the load resistor. The power switches and their antiparallel diodes (i.e., Q_1, D_1, Q_2, D_2, Q_3, D_3, Q_4, and D_4) form the two inverter legs. Note that the output rectifier can take the form of either a full-bridge or a double-current rectifier and that the resultant full-bridge converter operates in a similar fashion to that with a full-wave rectifier.

By controlling the four power switches Q_1 to Q_4, the voltage between points A and B, v_{AB}, can be made to be a high-frequency ac square voltage with an amplitude of V_{in}. Here, v_{AB} is transferred by the transformer T_r and then rectified by the output rectifier to give v_{rect}, which is a dc square-waveform voltage with an amplitude of V_{in}/K. Through the output filter, which consists of L_f and C_f, v_{rect} is filtered to a dc voltage V_o, which is the output voltage.

Soft-Switching PWM Full-Bridge Converters: Topologies, Control, and Design, First Edition. Xinbo Ruan.

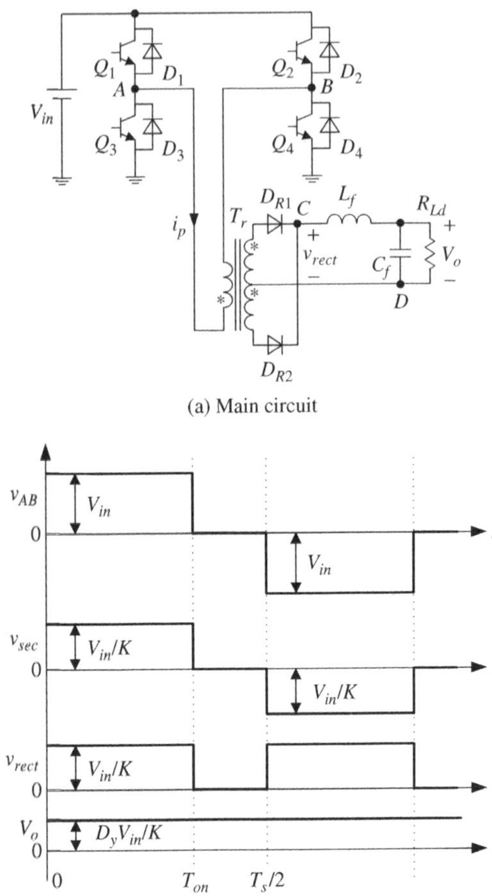

(a) Main circuit

(b) Key waveforms

Figure 2.1 Basic full-bridge converter

In order to obtain v_{rect}, a PWM dc square-waveform voltage, an ac square-wave voltage is required in the transformer secondary windings. This can be achieved by adopting a basic modulation strategy, as shown in Figure 2.2. The two switches in the diagonal line of the two bridge legs turn on/off simultaneously. The on-time T_{on} of each switch is determined by the duty cycle D_y of v_{rect}; that is, $T_{on} = D_y \cdot T_s/2$, where T_s is the switching period.

Using the basic PWM strategy, we can readily obtain the waveform of v_{rect}. Specifically, we can (i) keep the on-time of Q_2 and Q_4 unchanged, adjust the turn-on instant of Q_1 and Q_3 forward, or make the on-time equal to $T_s/2$; or (ii) keep the on-time of Q_1 and Q_3 unchanged, adjust the turn-off instant of Q_2 and Q_4 backward, or make the on-time equal to $T_s/2$; or (iii) push the turn-on instant of Q_1 and Q_3 forward or make the on-time equal to $T_s/2$ and at the same time push the turn-off instant of Q_2 and Q_4 backward or make the on-time equal to $T_s/2$. It should be noted that v_{AB} is the

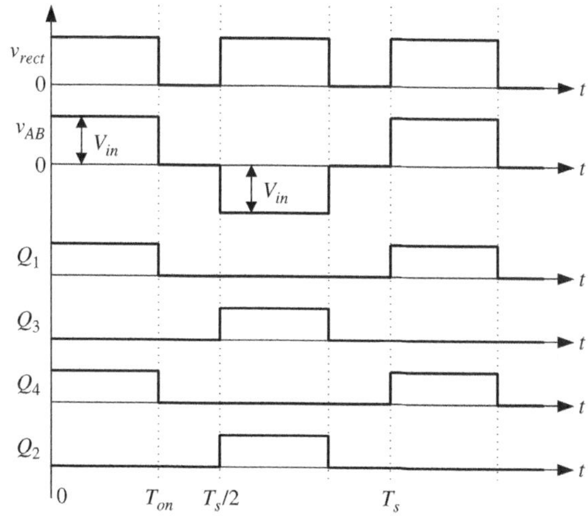

Figure 2.2 Basic PWM strategy

same as in the basic modulation strategy because only when Q_1 and Q_4 are turned on simultaneously, $v_{AB} = +V_{in}$; and only when Q_2 and Q_3 are turned on simultaneously, $v_{AB} = -V_{in}$. Thus, keeping the overlap on-time of the two diagonal power switches unchanged (i.e., $T_{on(overlap)} = T_{on}$), adjusting the turn-on instant of the power switches forward, or adjusting the turn-off instant of the power switches backward will not affect v_{AB}. A family of PWM strategies can thus be obtained for the full-bridge converter, as shown in Figure 2.3. All previously proposed PWM strategies belong to this family of modulation strategies.

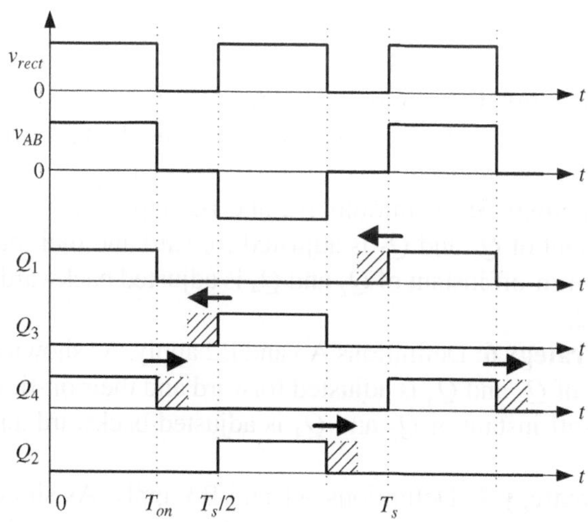

Figure 2.3 PWM strategies for obtaining the same waveform of v_{AB}

2.1.2 *Definition of On-Time of Power Switches*

According to the length of additional on-time T_{add}, the on-time of each bridge leg has three possible adjustments: it can remain unchanged, it can be increased but to less than $T_s/2$, or it can be increased to $T_s/2$.

1. **Definition of On-Time of Q_1 and Q_3:**
   a. A1: the on-time remains unchanged; that is, $T_{on} = D_y \cdot T_s/2$;
   b. A2: the turn-on instant is adjusted forward and the on-time is increased to a value less than $T_s/2$;
   c. A3: the turn-on instant is adjusted forward and the on-time is increased to $T_s/2$.
2. **Definition of On-Time of Q_2 and Q_4:**
   a. B1: the on-time remains unchanged; that is, $T_{on} = D_y \cdot T_s/2$;
   b. B2: the turn-off instant is adjusted backward and the on-time is increased to a value less than $T_s/2$;
   c. B3: the turn-on instant is adjusted backward and the on-time is increased to $T_s/2$.

2.1.3 *A Family of PWM Strategies*

According to the on-time of the two bridge legs, there are $3 \times 3 = 9$ possible PWM strategies, as shown in Figure 2.4. These are as follows:

- **Modulation strategy 1:** Definitions A1 and B1 apply. As shown in Figure 2.4a, the on-time of both bridge legs remains unchanged. This is the basic PWM strategy.
- **Modulation strategy 2:** Definitions A2 and B1 apply. As shown in Figure 2.4b, the turn-on instant of Q_1 and Q_3 is adjusted forward and their on-time is less than $T_s/2$, while the on-time of Q_2 and Q_4 remains unchanged.
- **Modulation strategy 3:** Definitions A3 and B1 apply. As shown in Figure 2.4c, the turn-on instant of Q_1 and Q_3 is adjusted forward and their on-time is equal to $T_s/2$, while the on-time of Q_2 and Q_4 remains unchanged.
- **Modulation strategy 4:** Definitions A1 and B2 apply. As shown in Figure 2.4d, the on-time of Q_1 and Q_3 remains unchanged, while the turn-off instant of Q_2 and Q_4 is adjusted backward and their on-time is less than $T_s/2$.
- **Modulation strategy 5:** Definitions A2 and B2 apply. As shown in Figure 2.4e, the turn-on instant of Q_1 and Q_3 is adjusted forward and their on-time is less than $T_s/2$, while the turn-off instant of Q_2 and Q_4 is adjusted backward and their on-time is less than $T_s/2$.
- **Modulation strategy 6:** Definitions A3 and B2 apply. As shown in Figure 2.4f, the turn-on instant of Q_1 and Q_3 is adjusted forward and their on-time is equal to $T_s/2$, while the turn-off instant of Q_2 and Q_4 is adjusted backward and their on-time is less than $T_s/2$.
- **Modulation strategy 7:** Definitions A1 and B3 apply. As shown in Figure 2.4g, the on-time of Q_1 and Q_3 remains unchanged, while the turn-off instant of Q_2 and Q_4 is adjusted backward and their on-time is equal to $T_s/2$.

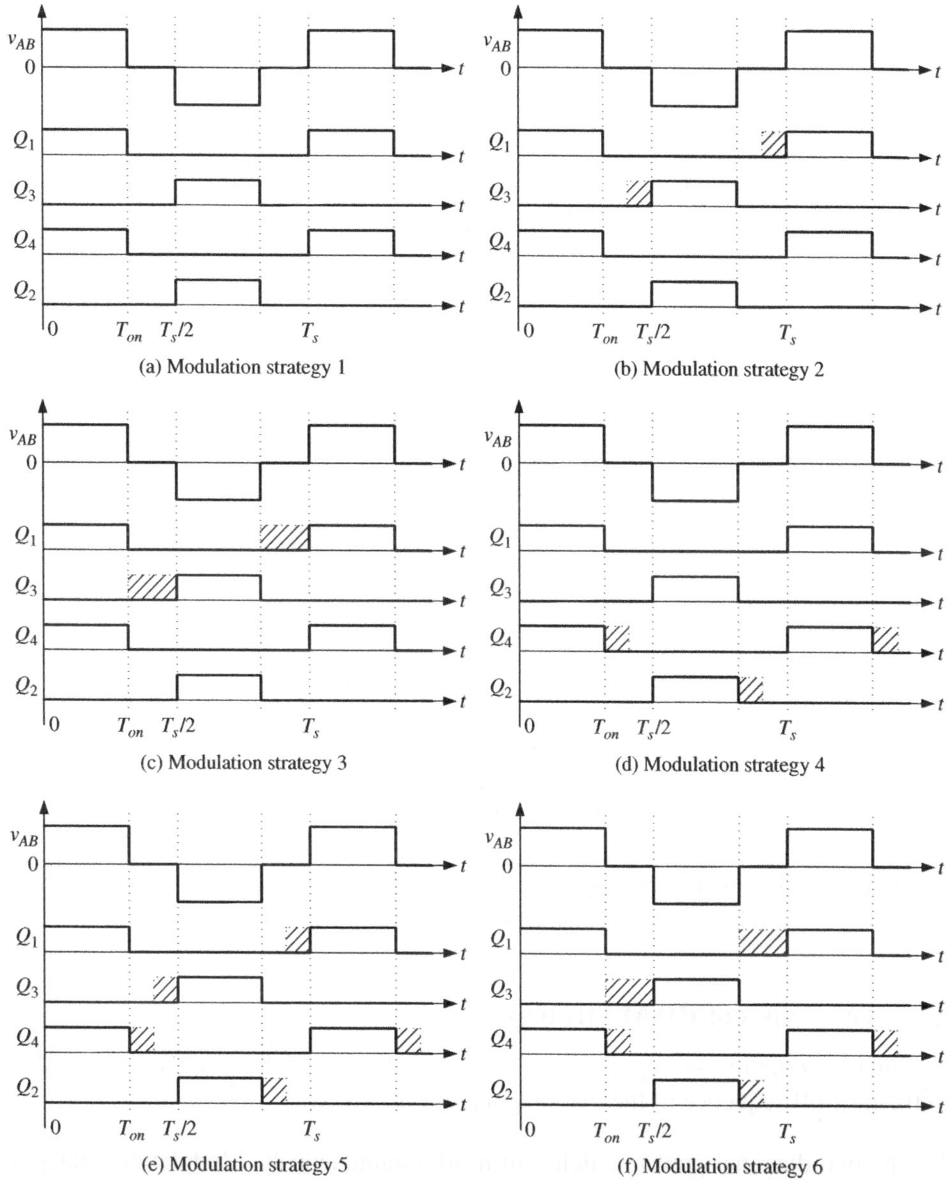

Figure 2.4 Family of PWM strategies for full-bridge converters

- **Modulation strategy 8:** Definitions A2 and B3 apply. As shown in Figure 2.4h, the turn-on instant of Q_1 and Q_3 is adjusted forward and their on-time is less than $T_s/2$, while the turn-off instant of Q_2 and Q_4 is adjusted backward and their on-time is equal to $T_s/2$.
- **Modulation strategy 9:** Definitions A3 and B3 apply. As shown in Figure 2.4i, the turn-on instant of Q_1 and Q_3 is adjusted forward and their on-time is equal to $T_s/2$,

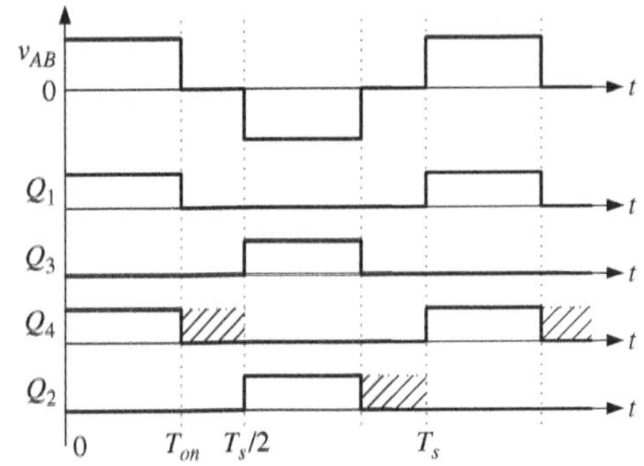

(g) Modulation strategy 7

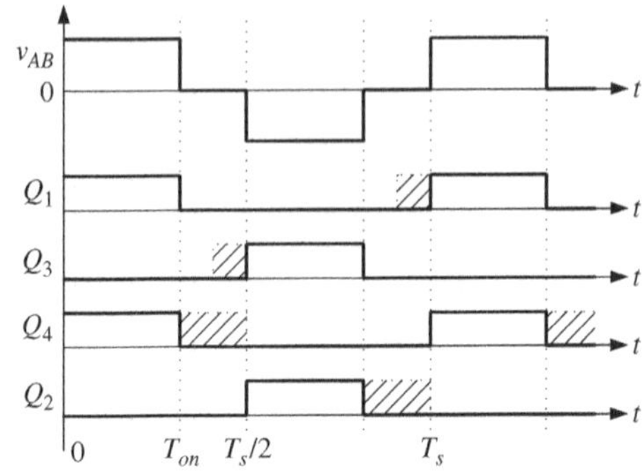

(h) Modulation strategy 8

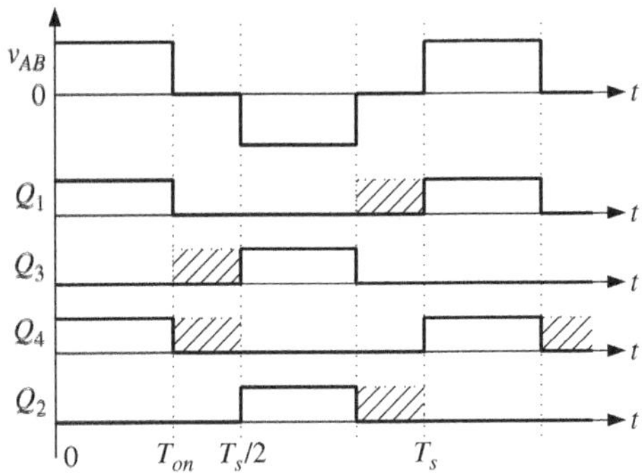

(i) Modulation strategy 9

Figure 2.4 (*Continued*)

while the turn-off instant of Q_2 and Q_4 is adjusted backward and their on-time is equal to $T_s/2$. This is the popular phase-shift modulation strategy.

2.2 Two Types of PWM Strategy

The nine modulation strategies can be categorized into two different types according to the turn-off sequence of the two diagonal power switches:

1. The two diagonal power switches turn off simultaneously. Modulation strategies 1–3 belong to this type.
2. The two diagonal power switches turn off in a staggered manner, one before the other. Modulation strategies 4–9 belong to this type.

With reference to Figure 2.1a, according to the on–off status of the four power switches, the full-bridge converter has three kinds of operating state:

- **+1 state:** Switches Q_1 and Q_4 conduct simultaneously. The voltage across points A and B, v_{AB}, is V_{in}; that is, $v_{AB} = V_{in}$.

- **Zero state:** Switches Q_1 (D_1) and Q_3 (D_3) conduct simultaneously or Q_2 (D_2) and Q_4 (D_4) conduct simultaneously. Here, $v_{AB} = 0 = (0)V_{in}$.
- **−1 state:** Switches Q_2 and Q_3 conduct simultaneously, giving $v_{AB} = -V_{in}$.

Thus, the full-bridge converter has three kinds of switching transition: +1/−1 (or −1/+1), +1/0 (or 0/+1), and −1/0 (or 0/−1).

2.2.1 *The Two Diagonal Power Switches Turn Off Simultaneously*

Figure 2.5 shows the main circuit and its relevant waveforms corresponding to the first kind of modulation strategy, where L_{lk} is the leakage inductor of the transformer. In order to realize soft turn-off for these power switches, capacitors C_1 to C_4 are connected in parallel with these switches. When Q_1 and Q_4 conduct simultaneously, $v_{AB} = V_{in}$, as shown in Figure 2.6a. As Q_1 and Q_4 turn off simultaneously at t_0, the primary current i_p charges C_1 and C_4 and discharges C_2 and C_3, as shown in

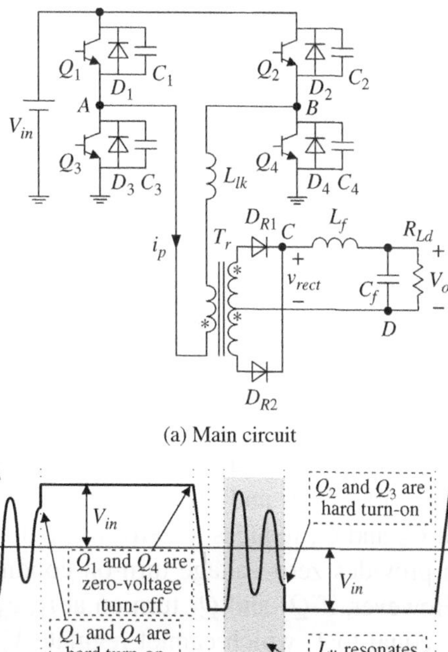

(a) Main circuit

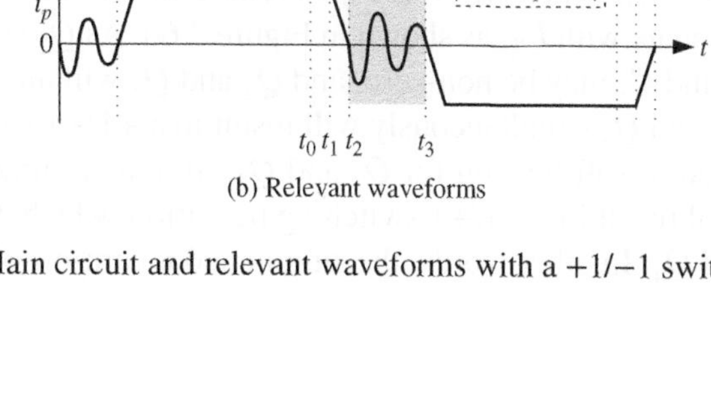

(b) Relevant waveforms

Figure 2.5 Main circuit and relevant waveforms with a +1/−1 switching transition

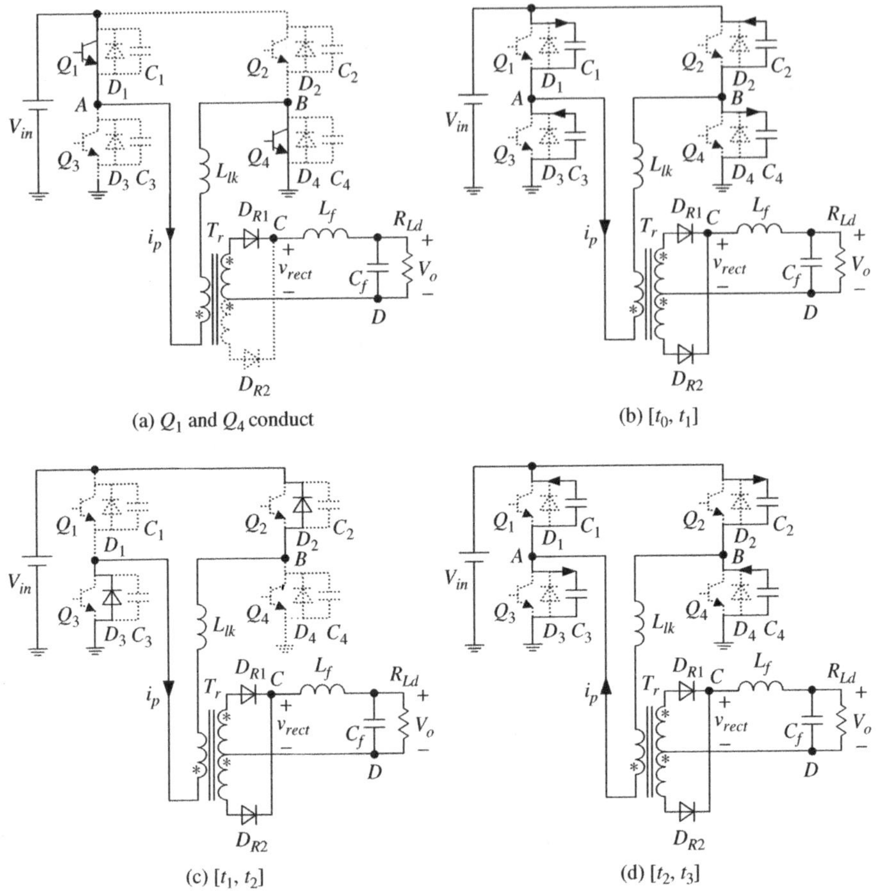

Figure 2.6 Equivalent circuit of the switching modes during the $[t_0, t_3]$ with +1/−1 switching transition

Figure 2.6b. Capacitors C_1 to C_4 limit the rise rate of the voltage of Q_1 and Q_4, allowing zero-voltage turn-off for Q_1 and Q_4. At t_1, the voltage of C_1 and C_4 rises to V_{in} and the voltage of C_2 and C_3 reduces to zero, while the antiparallel diodes D_2 and D_3 conduct, which provides zero-voltage turn-on condition for Q_2 and Q_3, as shown in Figure 2.6c. However, if Q_2 and Q_3 turn on at t_1, v_{AB} will be an ac square voltage with duty cycle equal to 1, which cannot regulate V_o. If Q_2 and Q_3 do not turn on at t_1, $v_{AB}=-V_{in}$, which will force i_p to reduce to zero at t_2. Thus, the parallel capacitors will resonate with L_{lk}, as shown in Figure 2.6d. As Q_2 and Q_3 turn on at t_3, the voltage of C_2 and C_3 may be non-zero, and Q_2 and Q_3 will undergo hard turn-on. So, turning off Q_1 and Q_4 simultaneously will result in a +1/−1 switching transition, which does not lead to soft turn-on for Q_2 and Q_3. Similarly, turning off Q_2 and Q_3 simultaneously will result in a −1/+1 switching transition, which does not allow soft turn-on for Q_1 and Q_4. In other words, the simultaneous turn-off of the two diagonal power switches will result in a +1/−1 or −1/+1 switching transition, which does not allow soft switching to be realized.

2.2.2 *The Two Diagonal Power Switches Turn Off in a Staggered Manner*

If the turn-off time instant of the two diagonal power switches is staggered (one switch turns off before the other), the switching transition will be improved. If Q_1 and Q_3 turn off before Q_4 and Q_2, respectively, we can define the bridge leg comprising Q_1 and Q_3 as the "leading leg" and the one comprising Q_2 and Q_4 as the "lagging leg."

2.2.2.1 Realization of Soft Switching for the Leading Leg

With reference to Figure 2.7a, when Q_1 and Q_4 conduct, $v_{AB}=+V_{in}$. As Q_1 turns off first, the primary current i_p charges C_1 and discharges C_3, as shown in Figure 2.7b. As a result, Q_1 undergoes zero-voltage turn-off, since C_1 and C_3 limit the rising rate of the voltage of Q_1. When the voltage of C_1 rises to V_{in}, the voltage of C_3 decays to zero and the antiparallel diode D_3 conducts naturally, providing a zero-voltage turn-on condition for Q_3. At this time, $v_{AB}=0$. Thus, this converter operates in +1/0 switching transition. Similarly, when Q_3 turns off, the converter operates in −1/0 switching transition, Q_3 undergoes zero-voltage turn-off, and Q_1 undergoes zero-voltage turn-on. In other words, the leading leg can realize zero-voltage switching (ZVS).

When the leading-leg switch turns off, the output filter inductor is in series with the leakage inductor and the primary current remains constant. Therefore, the leading leg can only realize ZVS and is easy to realize ZVS, because the energies of both the leakage inductor and the output filter inductor are used.

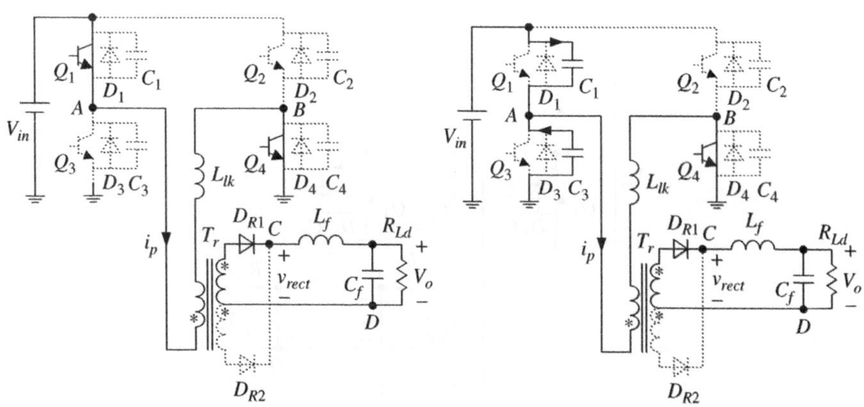

(a) Q_1 and Q_4 conduct simultaneously (b) Q_1 is turned off and Q_4 continues conducting

Figure 2.7 +1/0 switching transition

2.2.2.2 Realization of Soft Switching for the Lagging Leg

Zero State

Figure 2.8 shows the zero state of the full-bridge converter, where Q_3 and Q_4 are conducting and $v_{AB} = 0$. In fact, the primary current is flowing through D_3 and Q_4. If certain devices and/or elements are added to the converter, zero state can operate in two modes: the current constant mode, in which the primary current i_p is kept constant, and the current-reset mode, in which i_p is forced to zero and remains there.

ZVS for Lagging Leg

Figure 2.8 shows the zero state, which operates in current constant mode. The primary current i_p flows through D_3 and Q_4. When Q_4 is turned off, i_p charges C_4 and discharges C_2, as shown in Figure 2.9. Q_4 undergoes zero-voltage turn-off, thanks to the presence of C_2 and C_4. After Q_4 is turned off, the voltage of C_4 increases and

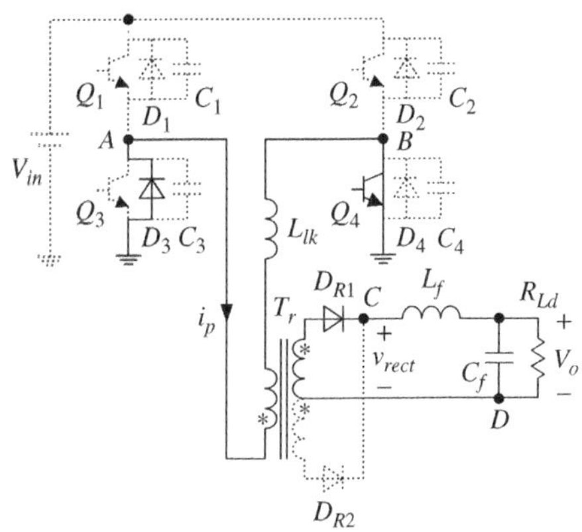

Figure 2.8 Zero state

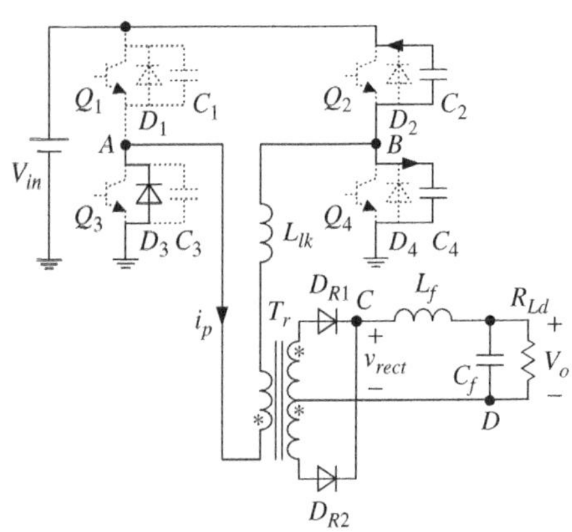

Figure 2.9 0/−1 switching transition

$v_{AB} = -v_{C4}$. This negative voltage makes i_p decay and the top secondary current decays accordingly; it is smaller than the output filter inductor current. Thus, the rectifier diode D_{R2} is forced to conduct and the bottom secondary winding provides the current shortfall. Since both of the rectifier diodes D_{R1} and D_{R2} conduct, the two secondary winding voltages are zero, and the primary winding voltage is zero accordingly. Thus, v_{AB} is fully applied to L_{lk}, making i_p decay. If the energy stored in the leakage inductor is sufficient, the voltage of C_4 will be charged to V_{in}, while the voltage of C_2 will reduce to zero. Thus, D_2 conducts naturally and Q_2 can undergo zero-voltage turn-on. At this time, $v_{AB} = -V_{in}$. If the energy stored in the leakage inductor is insufficient, the voltage of C_4 cannot be charged to V_{in} before i_p decays to zero and C_2 will not be completely discharged. After this, i_p flows in the negative direction, discharges C_4, and charges C_2. When Q_2 turns on, the voltage of C_2 is non-zero, and Q_2 undergoes hard turn-on. Regardless of whether C_2 is completely discharged or not, when Q_2 turns on (i.e., $v_{AB} = -V_{in}$), the full-bridge converter operates in 0/−1 switching transition.

Similarly, if the full-bridge converter operates in zero state when D_1 and Q_2 are conducting, the converter will operate in 0/+1 switching transition when Q_2 turns off. At the same time, Q_2 undergoes zero-voltage turn-off and Q_4 undergoes zero-voltage turn-on if the energy stored in the leakage inductor is adequate.

In conclusion, if the zero state operates in current constant mode, the lagging leg can realize ZVS. However, since only the energy stored in the leakage inductor is used to realize ZVS for the lagging leg, ZVS is more difficult to achieve there than in the leading leg. The lagging leg may fail to realize ZVS at light load.

ZCS for Lagging Leg

If zero state operates in current-reset mode, i_p is forced to decay to zero and then stays there. When Q_4 turns off, it is zero-current turn-off. As Q_2 turns on, L_{lk} limits the rising rate of i_p and Q_2 is zero-current turn-on. At this time, $v_{AB} = (-1)V_{in}$ and the full-bridge converter operates in 0/−1 switching transition. Similarly, when Q_2 turns off, the converter operates in 0/+1 switching transition, Q_2 undergoes zero-current turn-off, and Q_4 undergoes zero-current turn-on. In other words, in current-reset mode, the lagging leg can readily realize zero-current switching (ZCS).

Note that the power switches of the lagging leg cannot have a parallel capacitor, so as to ensure ZCS. Also note that during zero state, as i_p reduces to zero it should not continue to flow in reverse direction or else the lagging leg will fail to realize ZCS.

2.3 Classification of Soft-Switching PWM Full-Bridge Converters

According to the way in which soft switching is achieved in the leading and the lagging legs, the soft-switching PWM full-bridge converters can be categorized into two types: (i) ZVS PWM full-bridge converters, whose zero state operates in current constant mode. In this case, both the leading and the lagging leg realize ZVS. (ii) Zero-voltage-and-zero-current switching (ZVZCS) PWM full-bridge converters, whose zero state operates in current-reset mode. In this case, the leading leg realizes ZVS and the lagging leg ZCS.

2.4 Summary

This chapter systematically presented a family of PWM strategies for full-bridge converters. Two types of modulation strategies were identified according to the turn-off sequence of the two diagonal power switches. In the first, the two diagonal power switches turn off simultaneously and soft switching is not achieved. In the second, the two diagonal power switches turn off in a staggered manner (one before the other), introducing the concept of the leading leg and the lagging leg. The leading leg can only realize ZVS and can do so readily, while the lagging leg can realize ZVS and ZCS when the zero state operates in current constant mode and in current-reset mode, respectively. According to the soft switching type of the leading and lagging legs, the soft-switching PWM full-bridge converters can be further classified into ZVS and ZVZCS.

Reference

1. Ruan, X. and Yan, Y. (2000) Soft-switching techniques for PWM dc/dc full-bridge converters. Proceeding of the IEEE Power Electronics Specialist Conference, pp. 634–639.

3

Zero-Voltage-Switching PWM Full-Bridge Converters

Chapter 2 described the soft-switching techniques for pulse-width-modulation (PWM) full-bridge converters. In general, soft-switching PWM full-bridge converters can be categorized into two types: zero-voltage switching (ZVS) and zero-voltage-and-zero-current switching (ZVZCS). For ZVS PWM full-bridge converters, the zero state operates in current constant mode and both the leading leg and lagging leg realize ZVS. For ZVZCS PWM full-bridge converters, the zero state operates in current-reset mode and the leading leg realizes ZVS while the lagging leg realizes zero-current switching (ZCS). This chapter will present the topology and modulation strategies of ZVS PWM full-bridge converters. Taking the phase-shift modulation strategy as an example, the operating principle, conditions for achieving ZVS for the leading and lagging legs, and duty cycle loss of a ZVS PWM full-bridge converter will be analyzed.

3.1 Topologies and Modulation Strategies of ZVS PWM Full-Bridge Converters

3.1.1 Modulation of the Lagging Leg

For convenience of illustration, the main circuit of the full-bridge converter is redrawn in Figure 3.1, where the switches Q_1 and Q_3 form the leading leg and the switches Q_2 and Q_4 form the lagging leg. C_1 to C_4 are the parallel-connected capacitors of the switches for realization of ZVS. The lagging leg can realize zero-voltage turn-off naturally thanks to its parallel-connected capacitors. What concerns us here is the turn-on of the lagging leg.

Figure 3.2 shows the key waveforms of the primary current and drive signals of the four switches. Prior to t_0, Q_1 and Q_4 are conducting and the primary current i_p flows through them. At t_0, the leading-leg switch Q_1 is turned off with zero voltage and its antiparallel diode D_3 conducts. The full-bridge converter operates in zero state, $v_{AB} = 0$, and i_p flows through D_3 and Q_4. If the lagging-leg switch Q_4 turns off at t_x during $[t_0, t_1]$, i_p charges C_4 and discharges C_2, as shown in Figure 3.3, and begins

Soft-Switching PWM Full-Bridge Converters: Topologies, Control, and Design, First Edition. Xinbo Ruan.

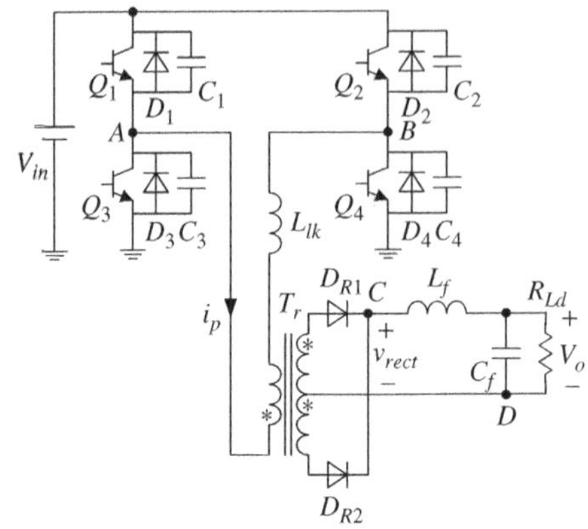

Figure 3.1 Main circuit of a full-bridge converter

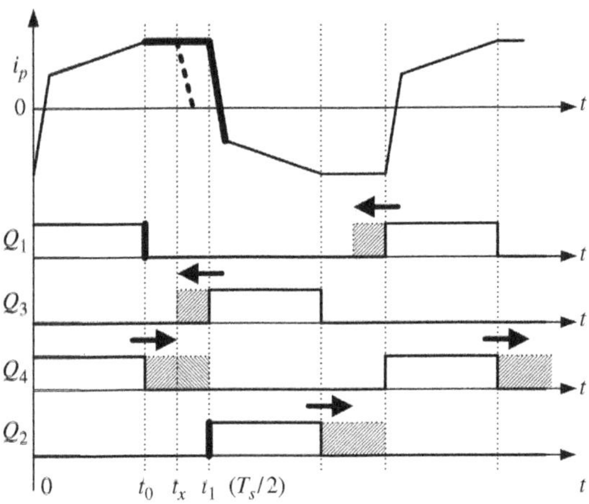

Figure 3.2 Modulation of the leading and the lagging leg to achieve ZVS

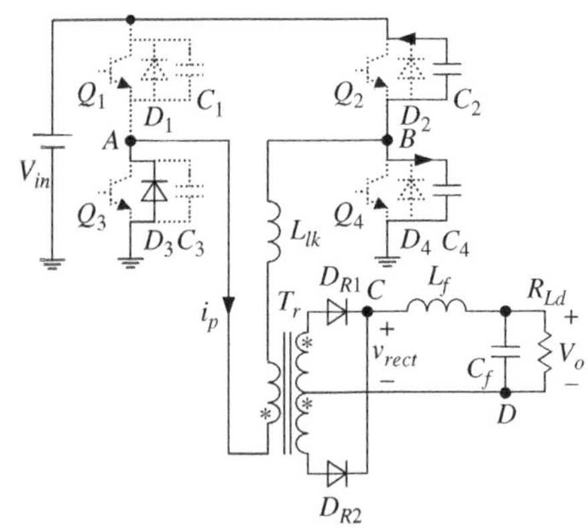

Figure 3.3 Switching of the lagging leg

to decay as given in the dashed line in Figure 3.2. When the voltage of C_4 rises to V_{in}, the voltage of C_2 decays to zero, the antiparallel diode D_2 conducts naturally, and $v_{AB} = -V_{in}$. At this point, i_p continues to decay to zero. If i_p reaches zero before t_1, from t_1 on C_2 will be charged and C_4 will be discharged, which makes Q_2 lose the condition of zero-voltage turn-on at t_1. The time it takes for i_p to decay to zero is dependent on the load: the lighter the load, the shorter the decay time. In order to ensure Q_2 will turn on with zero voltage at t_1 under any load, the turn-off instant of Q_4 should be pushed backward to t_1. The turn-off time instant of Q_2 is similar to that of Q_4. Therefore, the turn-off instant of the lagging-leg switches should be pushed backward to set the on-time to $T_s/2$.

3.1.2 Modulation of the Leading Leg

Like the lagging leg, the leading leg can realize zero-voltage turn-off naturally thanks to its parallel-connected capacitors. It is also important to examine its turn-on. As shown in Figure 3.2, the leading-leg switch Q_1 turns off with zero voltage at t_0, at which point the primary current i_p is flowing through the antiparalleled diode D_3, which provides the zero-voltage turn-on condition for Q_3. During $[t_0, t_1]$, the full-bridge converter operates in zero state, as shown in Figure 3.4. For the purpose of achieving ZVS, the zero state is in current constant mode, which means that i_p is kept constant and flows through D_3. Thus, during $[t_0, t_1]$, even if Q_3 is turned on, there is no current flowing through it. Therefore, Q_3 has three possible kinds of on-time: (i) it can be unchanged, (ii) it can be pushed forward and increased so it is less than $T_s/2$, or (iii) it can be pushed forward and increased to $T_s/2$. Similarly, Q_1 has three possible kinds of on-time.

3.1.3 Modulation Strategies of the ZVS PWM Full-Bridge Converters

There are three kinds of on-time for the leading leg and only one for the lagging leg. Therefore, three kinds of modulation strategy can be identified for the ZVS PWM full-bridge converter: strategies 7, 8, and 9 (Chapter 2), which are shown in Figure 3.5.

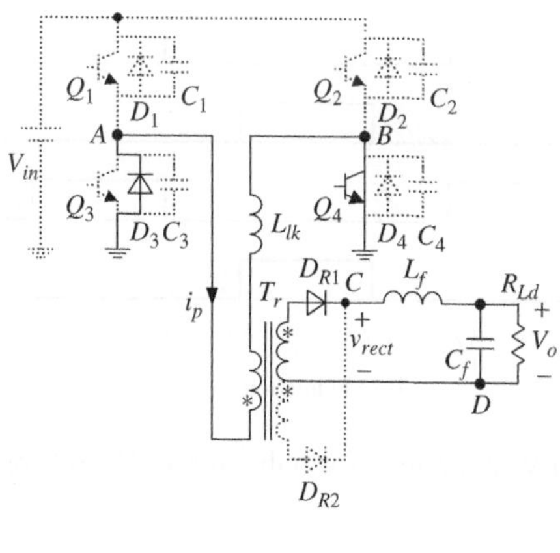

Figure 3.4 Zero state operating in current constant mode

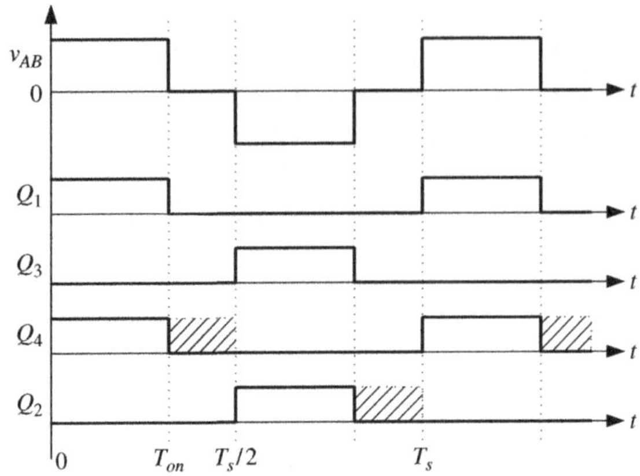

(a) Modulation strategy 7

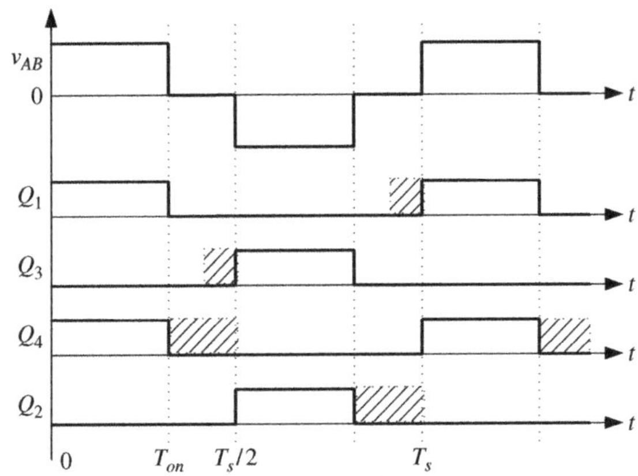

(b) Modulation strategy 8

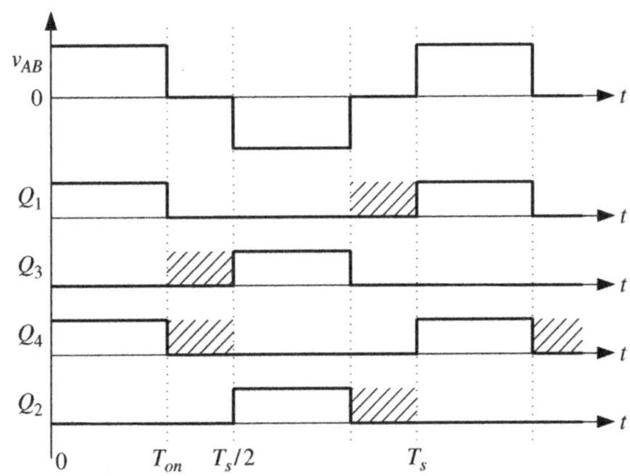

(c) Modulation strategy 9

Figure 3.5 PWM strategies suitable for a ZVS full-bridge converter

3.2 Operating Principle of ZVS PWM Full-Bridge Converter

Although there are three kinds of modulation strategy for ZVS PWM full-bridge converters, their operating principles are basically the same. In this section, the operating principle is analyzed through the example of the popular phase-shift modulation strategy.

Figure 3.6 shows the main circuit and key waveforms of the phase-shift-controlled full-bridge converter, where Q_1 to Q_4 are the power switches, D_1 to D_4 are the antiparallel diodes (or body diodes if Q_1 to Q_4 use metal-oxide-semiconductor field-effect

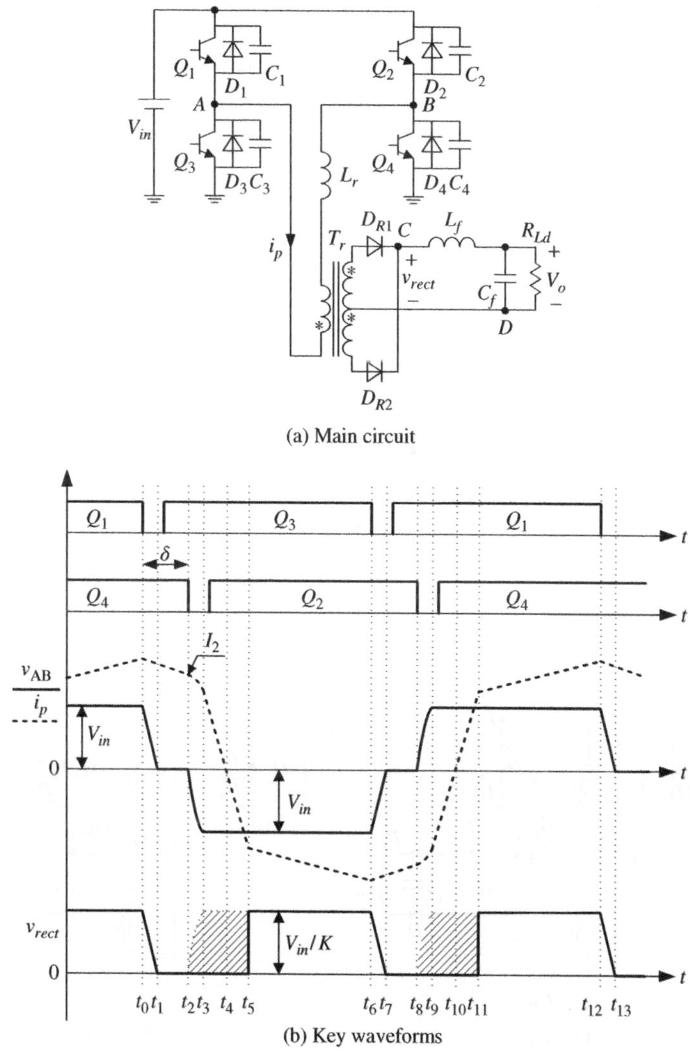

Figure 3.6 Main circuit and key waveforms of a phase-shift-controlled full-bridge converter

transistors (MOSFETs)) of Q_1 to Q_4, C_1 to C_4 are the snubber capacitors (or intrinsic capacitors) of Q_1 to Q_4, and L_r is the resonant inductor, including the transformer leakage inductor. The two switches of each bridge leg are operated in a complementary manner with 50% duty cycle and a phase shift is introduced between them to regulate the output voltage. The drive signals of Q_1 and Q_3 lead to those of Q_4 and Q_2, respectively, so the bridge leg consisting of Q_1 and Q_3 is defined as the leading leg and the one consisting of Q_2 and Q_4 as the lagging leg. In Figure 3.6b, the phase corresponding to the time interval $[t_0, t_2]$ is the phase shift δ, expressed as $\delta = \frac{t_2 - t_0}{T_s/2} \cdot 180°$. The smaller the phase shift δ, the higher the output voltage, and vice versa [1].

There are 12 switching modes for the ZVS PWM full-bridge converter in a switching period; their equivalent circuits are shown in Figure 3.7. Before beginning our analysis, we make the following assumptions: (i) all power devices and diodes are

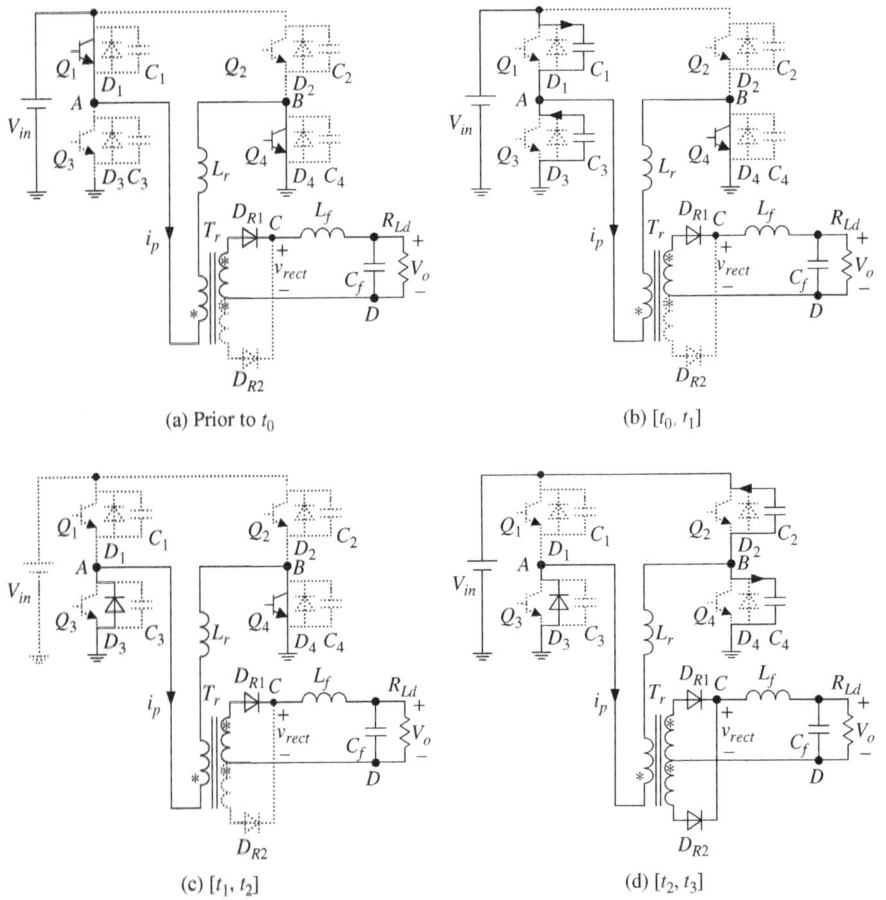

(a) Prior to t_0 (b) $[t_0, t_1]$ (c) $[t_1, t_2]$ (d) $[t_2, t_3]$

Figure 3.7 Equivalent circuits of switching modes

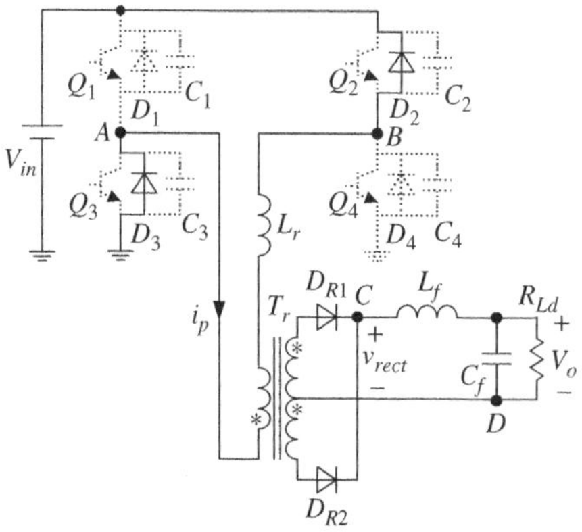

(e) $[t_3, t_4]$

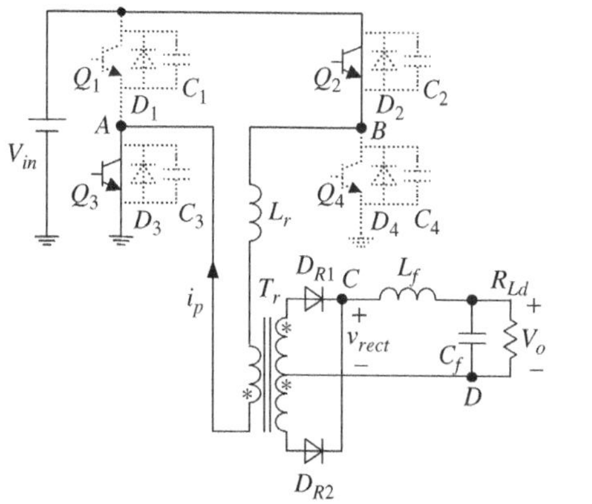

(f) $[t_4, t_5]$

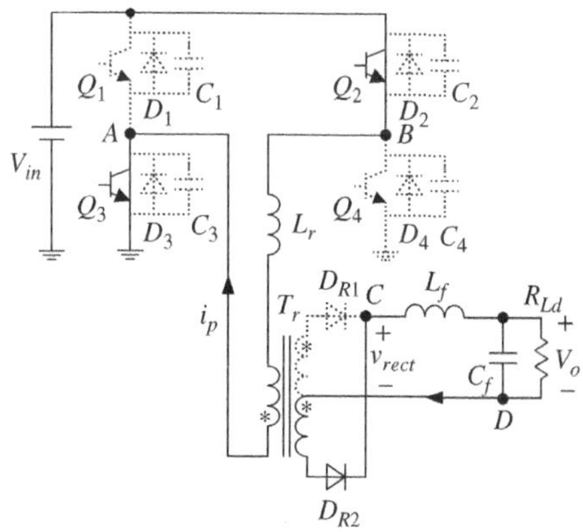

(g) $[t_5, t_6]$

Figure 3.7 (*Continued*)

ideal; (ii) all the capacitors, inductors, and transformer are ideal; (iii) $C_1 = C_3 = C_{lead}$, $C_2 = C_4 = C_{lag}$; and (iv) the resonant inductor is much smaller than the output filter inductor reflected to the primary side; that is, $L_r \ll K^2 \cdot L_f$, where K is the primary-to-secondary-windings-turns ratio.

1. **Mode 0 [at t_0] (Figure 3.7a):** Prior to t_0, Q_1 and Q_4 are conducting and the primary current i_p is flowing through them. In the secondary side, the rectifier diode D_{R1} is conducting.
2. **Mode 1 [t_0, t_1] (Figure 3.7b):** At t_0, Q_1 is turned off and the primary current i_p charges C_1 and discharges C_3. C_1 and C_3 limit the rate of change of the voltage of C_1. Thus, Q_1 is turned off with zero voltage. During this interval, the resonant inductor is in series with the output filter inductor reflected to the primary side. Since the output filter inductor is quite large, i_p is kept nearly constant during

$[t_0, t_1]$. Thus, the primary current i_p and the voltages of capacitors C_1 and C_3 can be expressed as:

$$i_p(t) = I_p(t_0) \triangleq I_1 \tag{3.1}$$

$$v_{C1}(t) = \frac{I_1}{2C_{lead}}(t - t_0) \tag{3.2}$$

$$v_{C3}(t) = V_{in} - \frac{I_1}{2C_{lead}}(t - t_0) \tag{3.3}$$

At t_1, the voltage of C_1 increases to V_{in} and the voltage of C_3 decays to zero, making the antiparallel diode D_3 conduct. The time interval of mode 1 is:

$$t_{01} = 2C_{lead}V_{in}/I_1 \tag{3.4}$$

3. **Mode 2 [t_1, t_2] (Figure 3.7c):** At t_1, D_3 conducts, clamping the voltage across C_3 at zero and thus providing the zero-voltage turn-on condition for Q_3. In order to ensure zero-voltage turn-on for Q_3, the dead time between the drive signals of Q_1 and Q_3, $t_{d(lead)}$, should be larger than t_{01}; that is:

$$t_{d(lead)} > 2C_{lead}V_{in}/I_1 \tag{3.5}$$

During mode 2, the primary current i_p equals the output filter inductor current reflected to the primary side; that is:

$$i_p(t) = i_{Lf}(t)/K \tag{3.6}$$

At t_2, i_p decays to I_2.

4. **Mode 3 [t_2, t_3] (Figure 3.7d):** At t_2, Q_4 is turned off with zero voltage thanks to capacitors C_2 and C_4. The primary current i_p charges C_4 and discharges C_2. Since $v_{AB} = -v_{C4}$, this negative voltage forces i_p to decay. The top secondary winding current decays accordingly and is lower than the output filter inductor current. Thus, the rectifier diode D_{R2} conducts and the bottom secondary winding makes up for the current shortfall. So, both of the two rectifier diodes conduct, setting both of the secondary winding voltages to zero, and as a result the primary winding voltage is zero. This means that v_{AB} is fully applied on the resonant inductor. Actually, during mode 3, L_r resonates with C_2 and C_4. The primary current i_p and the voltages of capacitors C_2 and C_4 are given by:

$$i_p(t) = I_2 \cos \omega_1(t - t_2) \tag{3.7}$$

$$v_{C4}(t) = Z_1 I_2 \sin \omega_1(t - t_2) \tag{3.8}$$

$$v_{C2}(t) = V_{in} - Z_1 I_2 \sin \omega_1(t - t_2) \tag{3.9}$$

where $Z_1 = \sqrt{L_r/(2C_{lag})}$ and $\omega_1 = 1/\sqrt{2L_r C_{lag}}$. At t_3, the voltage of C_4 is charged to V_{in} and the voltage of C_2 decays to zero, making the antiparallel diode D_2 conduct. The time interval of mode 3 is:

$$t_{23} = \frac{1}{\omega_1}\sin^{-1}\frac{V_{in}}{Z_1 I_2} \tag{3.10}$$

5. **Mode 4 [t_3, t_4] (Figure 3.7e):** At t_3, D_2 conducts and clamps the voltage of C_2 at zero, providing the zero-voltage turn-on condition for Q_2. The dead time between the drive signals of Q_2 and Q_4, $t_{d(lag)}$, should be larger than t_{23}; that is:

$$t_{d(lag)} > \frac{1}{\omega_1}\sin^{-1}\frac{V_{in}}{Z_1 I_2} \tag{3.11}$$

During this interval, the primary current i_p flows through D_2 and D_3 and the energy stored in the resonant inductor is regenerated to the input dc voltage source. The two rectifier diodes continue conducting, and both the primary and secondary winding voltages are zero. Hence, the input voltage is negatively applied on the resonant inductor, forcing i_p to decay linearly; that is:

$$i_p(t) = I_p(t_3) - \frac{V_{in}}{L_r}(t - t_3) \tag{3.12}$$

At t_4, i_p decays to zero and D_2 and D_3 are turned off naturally. The time interval of mode 4 is:

$$t_{34} = L_r I_p(t_3)/V_{in} \tag{3.13}$$

6. **Mode 5 [t_4, t_5] (Figure 3.7f):** At t_4, the primary current i_p crosses zero and continues increasing in the reverse direction through Q_2 and Q_3. Since i_p is still insufficient to power the load, the two rectifier diodes continue conducting and the primary and secondary winding voltages both remain at zero. Thus, the input voltage is applied on the resonant inductor in reverse polarity, forcing i_p to increase in reverse direction linearly; that is:

$$i_p(t) = -\frac{V_{in}}{L_r}(t - t_4) \tag{3.14}$$

At t_5, i_p increases to the value of the output filter inductor current reflected to the primary side, which equals $-I_{Lf}(t_5)/K$, the rectifier diode D_{R1} turns off, and the output inductor current flows through D_{R2}. The time interval of mode 5 is:

$$t_{45} = \frac{L_r I_{Lf}(t_5)/K}{V_{in}} \tag{3.15}$$

7. **Mode 6 [t_5, t_6] (Figure 3.7g):** From t_5, the input voltage source powers the load and the primary current i_p is:

$$i_p(t) = -\frac{V_{in} - KV_o}{L_r + K^2 L_f}(t - t_5) - \frac{I_{Lf}(t_5)}{K} \tag{3.16}$$

Since $L_r \ll K^2 \cdot L_f$, Equation 3.16 can be simplified to:

$$i_p(t) = -\frac{\frac{V_{in}}{K} - V_o}{KL_f}(t - t_5) - \frac{I_{Lf}(t_5)}{K} \tag{3.17}$$

At t_6, Q_3 is turned off, starting the second half-switching period [t_6, t_{12}], which is similar to the first [t_0, t_6].

3.3 ZVS Achievement of Leading and Lagging Legs

3.3.1 Condition for Achieving ZVS

From the analysis presented in Section 3.2, it can be seen that in order to achieve ZVS for the switches, the capacitor in parallel with the incoming switch should be completely discharged and the capacitor in parallel with the outgoing switch in the same bridge leg should be fully charged. If the intrinsic winding capacitor of the transformer C_{TR} is considered, it too should be completely discharged. Thus, the energy required for ZVS is:

$$E > \frac{1}{2}C_i V_{in}^2 + \frac{1}{2}C_i V_{in}^2 + \frac{1}{2}C_{TR} V_{in}^2 = C_i V_{in}^2 + \frac{1}{2}C_{TR} V_{in}^2 \ (i = lead, lag) \tag{3.18}$$

3.3.2 Condition for Achieving ZVS for the Leading Leg

As shown in Figure 3.7b, during the switching transition of the leading leg, the output filter inductor is reflected to the primary side and is in series with the resonant inductor. This means that the energy stored in both the output filter inductor and the resonant inductor is used to achieve ZVS for the leading leg. Generally, the output filter inductor is designed to be quite large, in order to reduce its current ripple, and the energy stored in it is sufficient to achieve ZVS for the leading leg even at light load.

3.3.3 Condition for Achieving ZVS for the Lagging Leg

As shown in Figure 3.7d, during the switching transition of the lagging leg, two rectifier diodes are both conducting. This makes the output filter inductor current freewheel through the two secondary windings, and it is not reflected to the primary side. Thus, it is only the energy stored in the resonant inductor that charges/discharges the capacitors in parallel with the switches of the lagging leg and discharges the intrinsic winding capacitor. In order to realize ZVS for the lagging leg, the energy stored in the resonant inductor should meet the requirement as:

$$\frac{1}{2}L_r I_2^2 > C_{lag} V_{in}^2 + \frac{1}{2}C_{TR} V_{in}^2 \tag{3.19}$$

Since the output filter inductor does not participate in achieving ZVS for the lagging leg and the resonant inductor is much smaller than the output filter inductor reflected to the primary side, it is much more difficult to realize ZVS in the lagging leg than in the leading leg, and it will fail to realize ZVS at light load. From Equation 3.19, it can be seen that in order to ensure the lagging leg achieves ZVS at light load, we can enlarge the resonant inductor. For a given minimum load current required to achieve ZVS, the value of $I_{2\min}$ can be obtained, and hence the required resonant inductor can be calculated according to Equation 3.19.

3.4 Secondary Duty Cycle Loss

As seen in Figure 3.6, during the intervals $[t_2, t_5]$ and $[t_8, t_{11}]$, $v_{AB} = V_{in}$ and $v_{AB} = -V_{in}$, respectively, and the primary current i_p changes its polarity. Since i_p is smaller than the output filter inductor current reflected to the primary side, both the rectifier diodes conduct, making the rectified voltage v_{rect} vanish. This means that the secondary side loses the input voltage reflected to the secondary during $[t_2, t_5]$ and $[t_8, t_{11}]$. The ratio of the time interval of the lost voltage to the half-switching period is called "secondary duty cycle loss" D_{loss} and is given by:

$$D_{loss} = \frac{t_{25}}{T_s/2} \tag{3.20}$$

Since t_{23} is very short, it can be ignored. t_{35} is derived as:

$$t_{35} = \frac{L_r[I_2 + I_{Lf}(t_5)/K]}{V_{in}} \tag{3.21}$$

Meanwhile, the output filter inductor is assumed to be sufficiently large, so its current ripple can be ignored; that is, $I_{Lf}(t_5) = I_o$ and $I_2 = I_o/K$. Substituting these into Equations 3.20 and 3.21 leads to:

$$D_{loss} = \frac{2L_r \cdot 2I_o/K}{T_s V_{in}} = \frac{4L_r I_o f_s}{K V_{in}} \tag{3.22}$$

From Equation 3.22, it can be concluded that (i) the larger the resonant inductor, the larger the duty cycle loss, (ii) the larger the load current, the larger the duty cycle loss, and (iii) the smaller the input voltage, the larger the duty cycle loss.

The duty cycle loss results in the reduction of the effective secondary duty cycle. In order to obtain the required output voltage, the primary-to-secondary-winding-turns ratio of the transformer, K, should be reduced. The reduced turns ratio causes two problems. First, the primary current increases, leading to increased conduction loss and increased current stress of the switches. Second, the voltage stress of the rectifier diodes is increased, requiring the use of the diode with a higher voltage rating, which has a higher forward voltage drop; thus the conduction loss of the rectifier diodes increases.

In order to reduce the resonant inductor while achieving ZVS in a wide load range, a saturable inductor can be used as the resonant inductor in place of a linear one [2].

3.5 Commutation of the Rectifier Diodes

As discussed in Section 3.2, during $[t_2, t_5]$ both the primary and secondary winding voltages are zero, due to the simultaneous conduction of both rectifier diodes. In this section, the commutation of the rectifier diodes is discussed. As pointed out in Chapter 1, the output rectifier circuit of the full-bridge converter can be a full-wave rectifier, a full-bridge rectifier, or a current-doubler rectifier. The full-wave rectifier is usually adopted when the output voltage is relatively low, the full-bridge rectifier when the output voltage is relatively high, and the current-doubler rectifier when the output voltage is relatively low but the output current is relatively large due to simple transformer configuration and ease of manufacture of the output filter inductors. In this section, the full-wave rectifier and full-bridge rectifier are chosen to illustrate the commutation of the rectifier diodes.

3.5.1 *Full-Bridge Rectifier*

During $[t_2, t_5]$, all the rectifier diodes conduct, clamping both the primary and the secondary winding voltages at zero. Thus, the primary current is independent of the secondary side and is determined by the input voltage and resonant inductor. Figure 3.8 shows the full-bridge rectifier circuit and the waveforms of the rectifier diode currents.

At t_2, the output filter inductor current flows through the rectifier diodes D_{R1} and D_{R4}. During $[t_2, t_5]$, the primary current i_p decays, and the secondary current i_s follows it. The secondary current is smaller than the output filter inductor current; that is, $i_s < i_{Lf}$. Hence, D_{R2} and D_{R3} begin to conduct, providing a flow path for the difference between i_s and i_{Lf}. So, we have:

$$i_{DR1} + i_{DR2} = i_{Lf} \tag{3.23}$$

$$i_{DR2} + i_s = i_{DR4} \tag{3.24}$$

Generally, the same type of diode is adopted for D_{R1} to D_{R4}. D_{R1} and D_{R4} work in the same manner, as do D_{R2} and D_{R3}; that is:

$$i_{DR1} = i_{DR4} \tag{3.25}$$

$$i_{DR2} = i_{DR3} \tag{3.26}$$

Neglecting the magnetizing current, the relationship between the primary and secondary currents is:

$$i_p = i_s/K \tag{3.27}$$

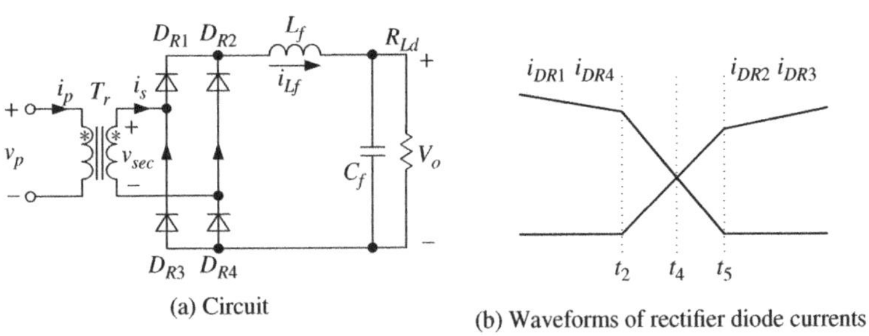

Figure 3.8 Commutation of a full-bridge rectifier

According to Equations 3.23–3.27, the four rectifier diode currents can be derived as:

$$i_{DR1} = i_{DR4} = \frac{1}{2}(i_{Lf} + Ki_p) \tag{3.28}$$

$$i_{DR2} = i_{DR3} = \frac{1}{2}(i_{Lf} - Ki_p) \tag{3.29}$$

From Equations 3.28 and 3.29, the commutation of the rectifier diodes can be illustrated as follows:

1. During $[t_2, t_4]$, $i_p > 0$, the currents in D_{R1} and D_{R4} are larger than the currents in D_{R2} and D_{R3}; that is:

$$i_{DR1} = i_{DR4} > i_{DR2} = i_{DR3} \tag{3.30}$$

2. At t_4, $i_p = 0$, all four rectifier diode currents are identical, being equal to half of the output filter inductor current; that is:

$$i_{DR1} = i_{DR4} = i_{DR2} = i_{DR3} = i_{Lf}/2 \tag{3.31}$$

3. During $[t_4, t_5]$, $i_p < 0$, the currents of D_{R1} and D_{R4} are lower than those of D_{R2} and D_{R3}; that is:

$$i_{DR1} = i_{DR4} < i_{DR2} = i_{DR3} \tag{3.32}$$

4. At t_5, $i_p = -i_{Lf}/K$, all of the output filter inductor current flows through D_{R2} and D_{R3}, while the currents of D_{R1} and D_{R4} are zero; that is:

$$i_{DR2} = i_{DR3} = i_{Lf} \tag{3.33}$$

$$i_{DR1} = i_{DR4} = 0 \tag{3.34}$$

So, D_{R1} and D_{R4} naturally turn off and D_{R2} and D_{R3} continue to conduct current to the output filter inductor. The commutation of the rectifier diodes is complete.

3.5.2 *Full-Wave Rectifier*

Figure 3.9 shows the full-wave rectifier circuit and the waveforms of the rectifier diode currents, from which we have:

$$i_{s1} = i_{DR1} \tag{3.35}$$

$$i_{s2} = -i_{DR2} \tag{3.36}$$

At t_2, the output filter inductor current flows through the rectifier diodes D_{R1}. During $[t_2, t_5]$, the primary current i_p decays and the top secondary winding current i_{s1} follows it. The top secondary winding current is smaller than the output filter inductor current; that is, $i_{s1} < i_{Lf}$. Thus, D_{R2} begins to conduct, and the bottom secondary winding provides the current deficit. So, we have:

$$i_{DR1} + i_{DR2} = i_{Lf} \tag{3.37}$$

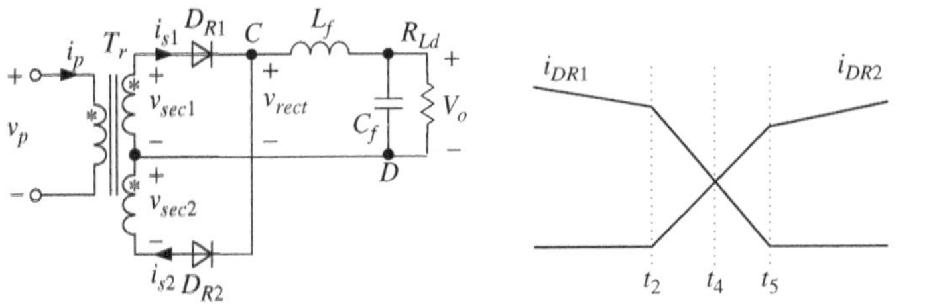

Figure 3.9 Commutation of a full-wave rectifier

Neglecting the magnetizing current, the relationship between the primary and secondary currents is:

$$i_{s1} + i_{s2} = Ki_p \tag{3.38}$$

According to Equations 3.35–3.38, the secondary currents and rectifier diode currents can be derived as:

$$i_{s1} = \frac{1}{2}(i_{Lf} + Ki_p) \tag{3.39}$$

$$i_{s2} = -\frac{1}{2}(i_{Lf} - Ki_p) \tag{3.40}$$

$$i_{DR1} = \frac{1}{2}(i_{Lf} + Ki_p) \tag{3.41}$$

$$i_{DR2} = \frac{1}{2}(i_{Lf} - Ki_p) \tag{3.42}$$

From Equations 3.41 and 3.42, the commutation of the rectifier diodes can be illustrated as follows:

1. During $[t_2, t_4]$, $i_p > 0$, the current of D_{R1} is larger than the current of D_{R2}; that is:

$$i_{DR1} > i_{DR2} \tag{3.43}$$

2. At t_4, $i_p = 0$, both of the rectifier diode currents are the same, being equal to half of the output filter inductor current; that is:

$$i_{DR1} = i_{DR2} = i_{Lf}/2 \tag{3.44}$$

3. During $[t_4, t_5]$, $i_p < 0$, the current of D_{R1} is lower than the current of D_{R2}; that is:

$$i_{DR1} < i_{DR2} \tag{3.45}$$

4. At t_5, $i_p = -i_{Lf}/K$, the output filter inductor current flows through D_{R2} and the current of D_{R1} is zero; that is:

$$i_{DR2} = i_{Lf} \tag{3.46}$$

$$i_{DR1} = 0 \tag{3.47}$$

So, D_{R1} naturally turns off and D_{R2} continues to conduct current through the output filter inductor. The commutation of the rectifier diodes is complete.

3.6 Simplified Design Procedure and Example

This section presents a simplified design procedure and example. The design of this converter involves complex interactions between circuit parameters and operating conditions [3–5]. Some assumptions and approximations are necessary.

The main specifications of the prototype are as follows:

- Input voltage $V_{in} = 210–373$ Vdc, which are obtained by a 220 Vac ± 20% through a diode rectifier and filter capacitor.
- Output voltage $V_o = 54$ Vdc.
- Output current $I_o = 10$ A.

3.6.1 *Turn Ratio of Transformer*

In order to fully utilize the transformer and reduce the current stress of the switches and the voltage stress of the rectifier diodes, the primary-to-secondary-windings-turns ratio of the transformer should be as large as possible. The full-bridge converter discussed in this book is a buck-derived converter, and the transformer turns ratio should be designed according to the minimum input voltage so that the required output voltage can be obtained over the entire input voltage range. Taking into account the duty cycle loss due to the resonant inductor, the maximum effective secondary duty cycle is set to 0.85 and the minimum magnitude of the secondary voltage is:

$$V_{\sec(\min)} = \frac{V_o + V_D + V_{Lf}}{D_{\sec(\max)}} = \frac{54 + 1.5 + 0.1}{0.85} = 65.41\ (\mathrm{V}) \tag{3.48}$$

where V_D is the forward conduction voltage drop of the rectifier diodes and V_{Lf} is the dc voltage drop in the output filter inductor.

The transformer turns ratio K is then 210/65.41 = 3.22. Here, the primary and secondary windings are selected as 18 and 6 turns, respectively, the actual transformer turns ratio is $K = 3$, and the maximum effective secondary duty cycle is:

$$D_{\sec(\max)} = \frac{V_o + V_D + V_{Lf}}{V_{in(\min)}/K} = \frac{54 + 1.5 + 0.1}{210.3/3} = 0.793 \tag{3.49}$$

3.6.2 *Resonant Inductor*

As discussed in Sections 3.3 and 3.4, in order to achieve ZVS for the lagging leg at light load, the value of the resonant inductor should preferably be large. However, a large resonant inductor results in a large duty cycle loss. There is thus a trade-off in the design of the resonant inductor. Here, the maximum effective secondary duty cycle is 0.793 and the maximum duty cycle loss can be selected as 0.15. According to Equation 3.22, the value of the resonant inductor is calculated as:

$$L_r = \frac{KV_{in(\min)}D_{loss}}{4I_{o\max}f_s} = \frac{3 \times 210.3 \times 0.15}{4 \times 10 \times 100 \times 10^3} = 23.66\ (\mu\text{H}) \tag{3.50}$$

Here, we choose $L_r = 24\ \mu$H.

3.6.3 *Output Filter Inductor and Capacitor*

The output filter inductance can be expressed as:

$$L_f = \frac{V_o}{(2f_s) \cdot \Delta I_{Lf}} \left(1 - \frac{V_o}{\frac{V_{in(\max)}}{K} - V_{Lf} - V_D} \right) \tag{3.51}$$

where ΔI_{Lf} is the output filter inductor current ripple. Substituting $V_{in(\max)} = 373$ V, $V_o = 54$ V, $f_s = 100$ kHz, $V_D = 1.5$ V, $V_{Lf} = 0.1$ V, $\Delta I_{Lf} = 0.2$, and $I_{o\max} = 2$ A into Equation 3.51, we get $L_f = 75.6\ \mu$H. Here, we choose $L_f = 75\ \mu$H.

The output filter capacitance can be expressed as:

$$C_f = \frac{V_o}{8L_f(2f_s)^2 \Delta V_{opp}} \left(1 - \frac{V_o}{\frac{V_{in(\max)}}{K} - V_{Lf} - V_D} \right) \tag{3.52}$$

Substituting $V_{in(\max)} = 373$ V, $V_o = 54$ V, $f_s = 100$ kHz, $V_D = 1.5$ V, $V_{Lf} = 0.1$ V, $L_f =$ 75 μH and $\Delta V_{opp} = 50$ mV into Equation 3.52, we obtain $C_f = 25.2\ \mu$F. There is an equivalent series resistor (ESR) in an electrolytic capacitor, and the output voltage ripple is mainly determined by the ripple voltage across the ESR resulting from the output filter inductor current ripple. The ESR should be limited to $\Delta V_{opp}/\Delta I_{Lf} =$ 50 mV/2 A = 25 mΩ. The relationship between the capacitance and ESR of an electrolytic capacitor is [6]:

$$C \cdot ESR = 60 \times 10^{-6} \tag{3.53}$$

Substituting ESR = 25 mΩ into Equation 3.53 leads to $C_f = 2400\ \mu$F. Here, two 1500 μF electrolytic capacitors are connected in parallel to realize the output filter capacitor.

3.6.4 *Power Devices*

The maximum input voltage is 373 V and the voltage stress of the switches is thus 373 V. The maximum current of the output filter inductor is $I_{o\max} + \frac{1}{2}\Delta I_{Lf} = 11$ A and the maximum current of the power switches is 11 A/K = 11 A/3 = 3.67 A. According to the voltage stress and maximum current of the switches, IRF840, whose voltage rating is 500 V and whose current rating is 8 A, is chosen for the primary side switches Q_1 to Q_4.

Since the output voltage is 54 V, the full-wave rectifier circuit is selected here, and the voltage stress of the rectifier diode is $2V_{in(\max)}/K = 2 \times 373\ \text{V}/3 = 248.67$ V. The rectifier diode suffers voltage oscillation and voltage spike due to the reverse recovery. Thus, the diode should have a voltage rating higher than 248.67 V × 2 = 497.3 V. To simplify the analysis, the rectifier diode can be considered to conduct during the half-switching period, and its root-mean-square current is:

$$I_{DR} = \frac{I_o}{\sqrt{2}} = \frac{10}{\sqrt{2}} = 7.07\ (\text{A}) \tag{3.54}$$

The maximum current flowing through the rectifier diodes is $i_{DR(\max)} = I_{o\max} + \frac{1}{2}\Delta I_{Lf} = 10 + \frac{1}{2} \times 2 = 11$ (A).

According to the calculated voltage and current stress, the fast recovery diode DESI12-06A, whose voltage rating is 600 V and current rating is 12 A, is selected as the rectifier diode.

3.6.5 *Load Range of ZVS*

The output capacitor of a MOSFET is nonlinear and can be replaced by a constant capacitor, expressed as [7]:

$$C_{lag} = \frac{4}{3} \times C_{oss\_25\text{V}} \times \sqrt{25/V_{in}} \tag{3.55}$$

where $C_{oss\_25\text{V}}$ is the output capacitor of the MOSFET when the drain-to-source voltage $V_{ds} = 25$ V.

3.6.5.1 Load Range of ZVS for the Lagging Leg

By neglecting the output filter inductor current ripple, we have $I_2 = I_o/K$. Thus, from Equations 3.55 and 3.19, and neglecting the transformer winding capacitor, the minimum load current for achieving ZVS for the lagging leg is:

$$I_{o\min\_\text{lag}} \geq KV_{in}\sqrt{\frac{40}{3}\frac{C_{oss\_25\text{V}}}{L_r\sqrt{V_{in}}}} \tag{3.56}$$

For IRF840, the $C_{oss\_25\text{V}}$ is 310 pF. Also, using Equation 3.56, and putting $K = 3$ and $L_r = 24$ μH, the minimum load current for achieving ZVS for the lagging leg over the entire input voltage range can be plotted as shown in the solid curve in Figure 3.10. As can be seen, the higher the input voltage, the larger the minimum load current required to achieve ZVS. In summary, the lagging leg can achieve ZVS at one-third full load over the entire input voltage range.

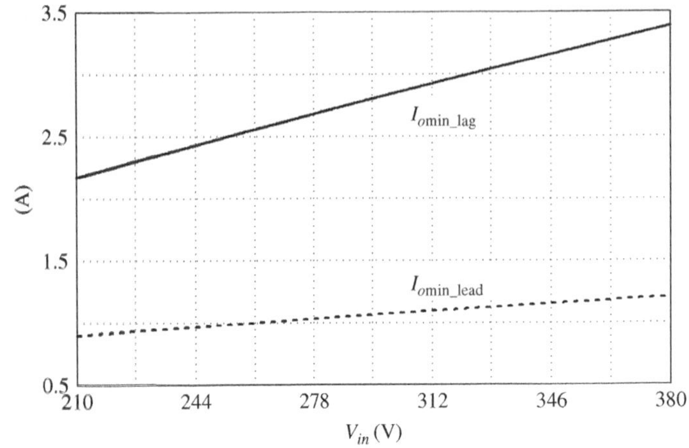

Figure 3.10 Minimum load current required to achieve ZVS for the lagging leg

3.6.5.2 Load Range of ZVS for the Leading Leg

During the switching transition of the leading leg, the parallel capacitors of the switches are charged/discharged in a linear manner due to the large output filter inductor. Similarly, by neglecting the output filter inductor current ripple, we have $I_1 = I_o/K$. So, according to Equation 3.4, the minimum load current required to achieve ZVS for the leading leg is:

$$I_{o\min\_lead} = \frac{40}{3} \frac{K\sqrt{V_{in}C_{oss\_25V}}}{t_{d(\text{lead})}} \tag{3.57}$$

where $t_{d(\text{lead})}$ is the dead time between the gate drive signals of the two leading-leg switches, set to 200 ns. Putting $t_{d(\text{lead})} = 200$ ns, $C_{oss\_25V} = 310$ pF, $K = 3$, and $L_r = 24$ μH into Equation 3.57, the minimum load current for achieving ZVS for the leading leg over the entire input voltage range can be plotted as shown in the dashed curve in Figure 3.10. As can be seen, the higher the input voltage, the larger the minimum load current required to achieve ZVS. In summary, the leading leg can achieve ZVS at 13% full load over the entire input voltage range.

3.7 Experimental Verification

In order to show the operating principle of the phase shift-controlled full-bridge converter, according to the parameter design presented in Section 3.6, experimental measurements from a 540 W prototype are presented here. The main power devices and elements are as follows:

- Q_1 (D_1) ~ Q_4 (D_4): IRF840;
- output rectifier diodes D_{R1} and D_{R2}: DESI12-06A;

- resonant inductor $L_r = 24\,\mu\text{H}$;
- transformer primary-to-secondary-winding-turns ratio $K = 18:6$;
- output filter inductor $L_f = 75\,\mu\text{H}$;
- output filter capacitor $C_f = 1500\,\mu\text{F} \times 2$; and
- switching frequency $f_s = 100\,\text{kHz}$.

Figure 3.11 shows the experimental waveforms under full load and nominal ac input voltage (220 Vac, 50 Hz). As can be seen, the waveforms of primary voltage v_{AB} and primary current i_p are very clean. The waveform of the secondary rectified voltage v_{rect} shows oscillatory behavior, which results from the reverse recovery of the rectifier diodes. It can be seen that there is a transition time for the primary current i_p to change its direction from positive (or negative) to negative (or positive), as marked by the dashed lines. This is the duty cycle loss.

Figures 3.12–3.14 shows the experimental waveforms of the drive signals, drain-to-source voltage, and drain current of the switches at full load, half load,

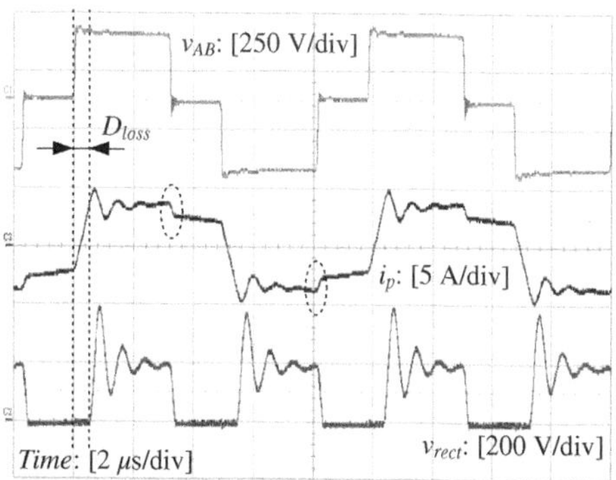

Figure 3.11 Experimental waveforms under full load and nominal ac input voltage

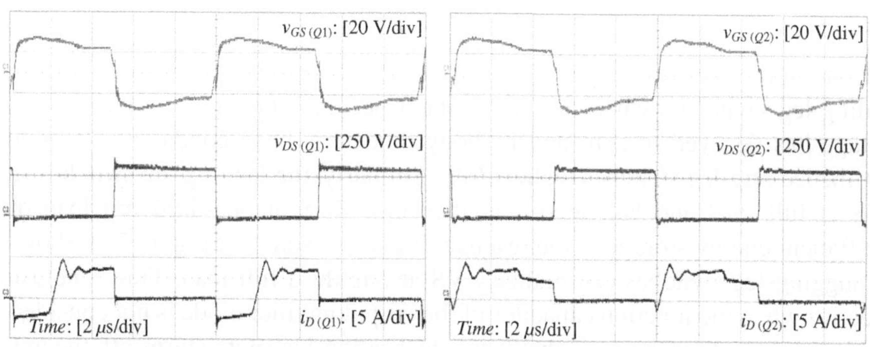

(a) Leading-leg switch Q_1 (b) Lagging-leg switch Q_2

Figure 3.12 Experimental waveforms of the drive signals, drain-to-source voltage, and drain current of switches at full load

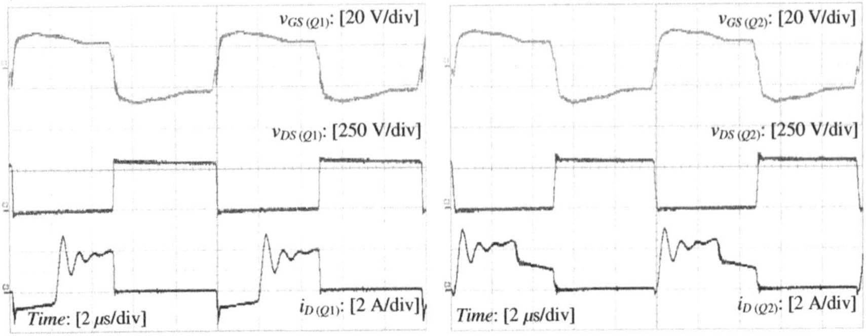

(a) Leading-leg switch Q_1 (b) Lagging-leg switch Q_2

Figure 3.13 Experimental waveforms of the drive signals, drain-to-source voltage, and drain current of switches at half load

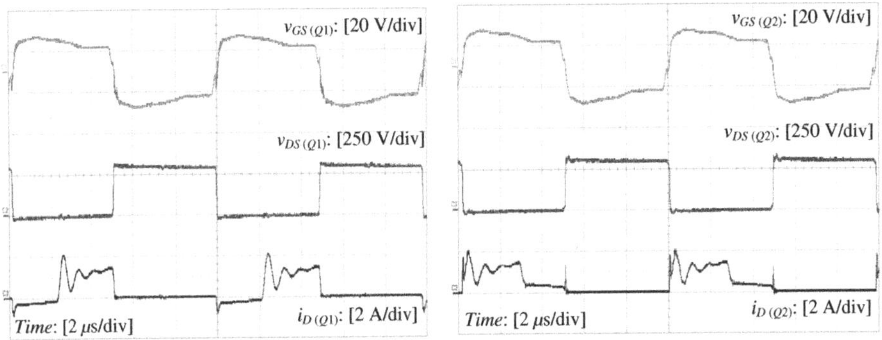

(a) Leading-leg switch Q_1 (b) Lagging-leg switch Q_2

Figure 3.14 Experimental waveforms of the drive signals, drain-to-source voltage, and drain current of switches at one-third full load

and one-third full load, respectively. From these figures, it can be seen that, for the leading-leg switches, when the drive signal becomes positive, the drain-to-source voltage has decayed to zero and its body diode has been conducting. This means that the leading-leg switches realize ZVS. Similarly, the lagging-leg switches realize ZVS at full load and half load, and they lose ZVS at one-third full load due to insufficient energy storage in the resonant inductor. Note that Figure 3.10 shows that the lagging-leg switches can realize ZVS at one-third full load. This conclusion is effective when the junction capacitor of the output rectifier diode is not considered. It will be explained in Chapter 6 that when the leading-leg switch turns off, the junction capacitor of the output rectifier diode participates in the resonance with the resonant inductor and the paralleled capacitors of leading-leg switches and the primary current step down from the output filter inductor current reflected to the primary side. This

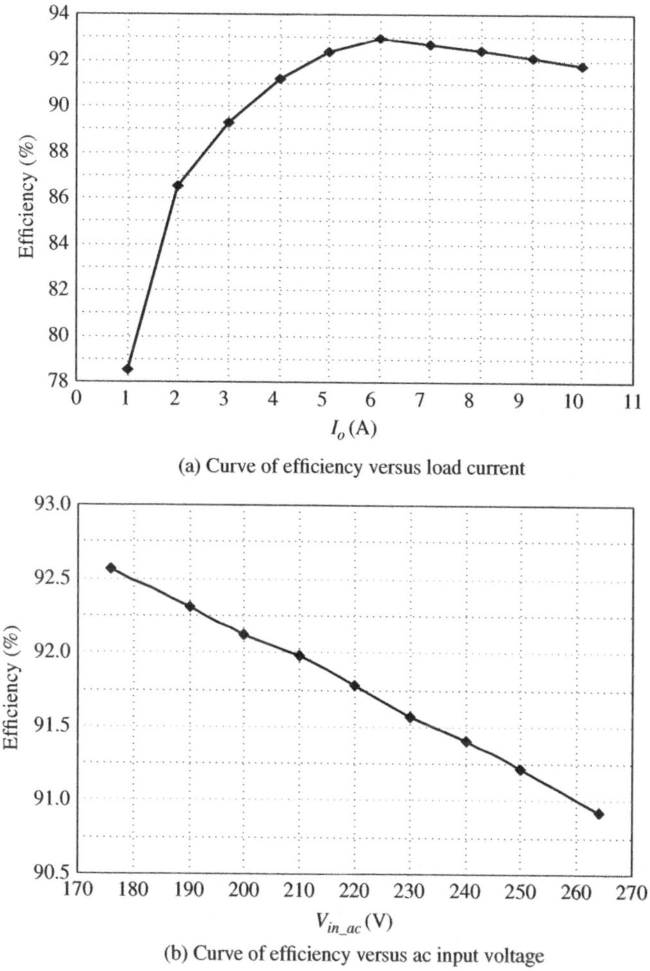

Figure 3.15 Overall conversion efficiency of the prototype of the full-bridge converter

can clearly be seen in Figure 3.11, as indicated in the dashed circles. So, due to the junction capacitor of the output rectifier diode, the practical load range of ZVS for the lagging leg is narrower than is shown in Figure 3.10.

Figure 3.15 shows the overall conversion efficiency of the prototype of the full-bridge converter. It should be noted that the measured efficiencies are from the single-phase ac voltage to the output, given consideration to the loss in the input diode rectifier. Figure 3.15a shows the measured efficiencies at different load currents under nominal input ac voltage (220 Vac, 50 Hz). As can be seen, the efficiency is highest at 6 A load current, and it is 93%. At full load, the efficiency is 91.8%. Figure 3.15b shows the measured efficiencies at full load over the input ac voltage range. It can be seen that the efficiency decays with the increase of the input ac voltage. This is because the full-bridge converter has zero state, in which no power is transferred to the load, but the primary current is freewheeling, leading to conduction loss in the power devices, resonant inductor, and transformer. The higher the input voltage, the longer the zero state and the larger the conduction loss, reducing the conversion efficiency.

3.8 Summary

This chapter discussed the PWM strategies for the ZVS full-bridge converter, selecting three suitable ones from the family proposed in Chapter 2. Taking the popular phase-shift modulation strategy as an example, the operating principle of the ZVS PWM full-bridge converter was presented. Realization of ZVS was discussed, and it was mentioned that ZVS is easier to realize with the leading leg than with the lagging. In the leading leg, the energy stored in both the resonant inductor and the output filter inductor is used to achieve ZVS, while in the lagging leg only the energy stored in the resonant inductor is used. The mechanism of secondary duty cycle loss due to the resonant inductor was revealed and the commutation of the output rectifier diodes in the full-bridge rectifier circuit and the full-wave rectifier circuit was analyzed. A simplified design procedure and an example of a ZVS PWM full-bridge converter were presented, and the experimental measurements from a 540 W prototype were provided.

References

1. Sable, D.M. and Lee, F.C. (1989) The operation of a full-bridge, zero-voltage-switched PWM converter. Proceedings of the Virginia Power Electronics Center Seminar, pp. 92–97.
2. Hua, G., Lee, F.C. and Jovanovic, M.M. (1991) An improved zero-voltage-switched PWM converter using a saturable inductor. Proceedings of the IEEE Power Electronics Specialist Conference, pp. 189–194.
3. Chen, Q., Loft, A.W. and Lee, F.C. (1991) Design trade-offs in 5-V output off-line zero-voltage, PWM converter. Proceedings of the IEEE International Telecommunication Energy Conference, pp. 616–623.
4. Sabate, J.A., Vlatkovic, V., Ridley, R.B. *et al.* (1990) Design considerations for high-voltage, high power full-bridge zero-voltage-switched PWM converter. Proceedings of the IEEE Applied Power Electronics Conference and Exposition, pp. 275–284.
5. Lofti, A.W., Sabate, J.A. and Lee, F.C. (1990) Design optimization of the zero-voltage-switched PWM converter. Proceedings of theVirginia Power Electronics Center Seminar, pp. 30–37.
6. Pressman, A.I., Billings, K. and Morey, T. (2009) *Switching Power Supply Design*, 2nd edn, McGraw-Hill Professional, New York.
7. Erickson, R.W. and Maksimovic, D. (2001) *Fundamentals of Power Electronics*, 2nd edn, Kluwer Academic Publishers, Boston, MA.

4

Zero-Voltage-Switching PWM Full-Bridge Converters with Auxiliary-Current-Source Networks

Chapter 3 presented the operating principle of the basic zero-voltage-switching (ZVS) pulse-width-modulation (PWM) full-bridge converters. The leading leg can readily realize ZVS because the energy stored in both the output filter inductor and the resonant inductor (including the transformer leakage inductor) is utilized to charge/discharge the parallel-connected capacitors of the leading-leg power switches. The lagging leg realizes ZVS using only the energy stored in the resonant inductor, which is much smaller than that in the output filter inductor. Thus, it is more difficult to realize ZVS in the lagging leg than in the leading leg, and the former will fail to realize ZVS at light load. Increasing the resonant inductor can expand the load range in order to achieve ZVS, but this will lead to an increase in the secondary duty cycle loss, resulting in a reduced effective duty cycle, and in order to obtain the desired output voltage, the primary-to-secondary-winding-turns ratio will have to be reduced. This leads to increased primary current and thus increased conduction loss, as well as increased voltage stress on the output rectifier diodes and thus increased voltage forward drop, reducing the conversion efficiency.

In order to avoid an increase of the resonant inductor while achieving ZVS in a wide load range, an auxiliary-current-source network can be incorporated into the basic ZVS PWM full-bridge converter. This chapter begins by describing the current-enhancement principle. During the switching transition of the leading or the lagging leg, the auxiliary current source and the transformer primary current simultaneously flow into/out of the midpoint of the bridge leg, thus both charging/discharging the parallel-connected capacitors. As a consequence, the power switches can realize ZVS even at light load. Next, an auxiliary-current-source network used to assist in ZVS, consisting of an inductor, two capacitors, and two diodes, is introduced, and

Soft-Switching PWM Full-Bridge Converters: Topologies, Control, and Design, First Edition. Xinbo Ruan.

its operating principle is analyzed. This network is added into the basic ZVS PWM full-bridge converter and connected with the lagging leg, allowing it to realize ZVS over the entire input voltage range and the entire load range, while reducing the duty cycle loss nearly to zero. The operating principle of the ZVS PWM full-bridge converter with the auxiliary current-source network is described and the parameter design is discussed. A 52.8 V/50 A output prototype is used to demonstrate the operating principle and parameter design of the ZVS PWM full-bridge converter. Finally, several auxiliary-current-source networks are presented, and their operating principles are discussed.

4.1 Current-Enhancement Principle

Figure 4.1 shows the main circuit and key waveforms of the basic ZVS PWM full-bridge converter. As can be seen in the figure, for the leading leg, consisting of

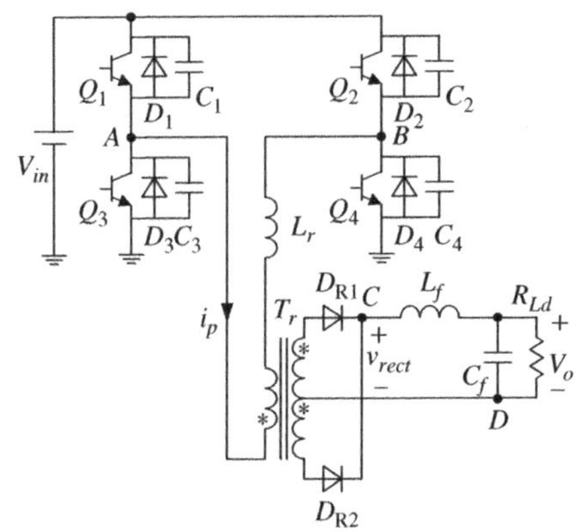

(a) Main circuit

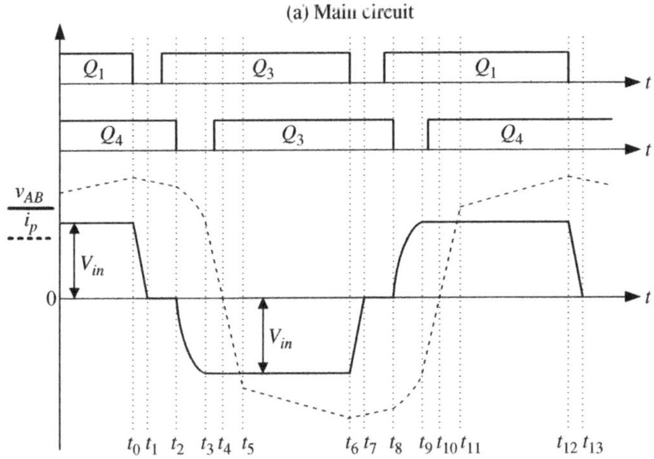

(b) Key waveforms

Figure 4.1 Basic ZVS PWM full-bridge converter

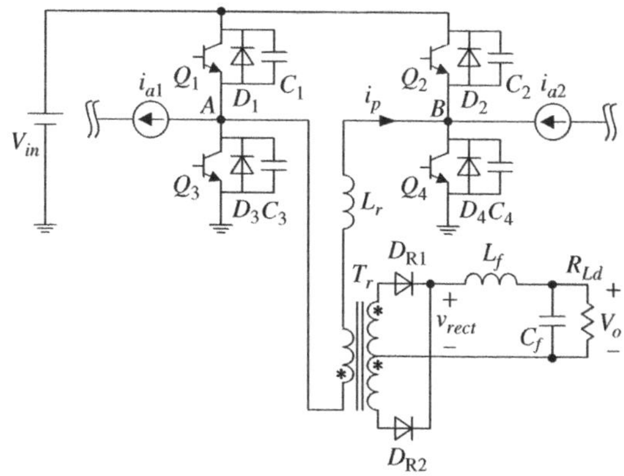

Figure 4.2 Auxiliary current source for use in achieving ZVS

Q_1 and Q_3, the primary current i_p flows into the midpoint A when Q_3 is turned off and out of point A when Q_1 is turned off; for the lagging leg, consisting of Q_2 and Q_4, i_p flows into the midpoint B when Q_4 is turned off and out of point B when Q_2 is turned off. If the auxiliary current sources i_{a1} and i_{a2} are connected to points A and B, respectively, during the switching transition of the leading leg, as shown in Figure 4.2, i_p and i_{a1} flow into/out of point A simultaneously. This means that the current flow into/out of point A is enhanced, and both currents i_p and i_{a1} charge/discharge the parallel capacitors of the leading-leg switches, thus allowing the leading leg to realize ZVS in a wide load range. Likewise, during the switching transition of the lagging leg, i_p and i_{a2} flow into/out of point B simultaneously, and the current flow into/out of point B is enhanced. Both currents i_p and i_{a2} charge/discharge the parallel capacitors of the lagging-leg switches, leading to a wide load range in which ZVS can be achieved for the lagging leg. As illustrated in Chapter 3, it is easy for the leading leg to realize ZVS because the energy stored in both the output filter inductor and the resonant inductor is utilized, so it is not necessary to add an auxiliary current source to point A.

4.2 Auxiliary-Current-Source Network

There are various methods for realizing the auxiliary current source. Figure 4.3a shows an auxiliary-current-source network [1], which comprises auxiliary inductor L_a, auxiliary capacitors C_{a1} and C_{a2}, and auxiliary diodes D_{a1} and D_{a2}. C_{a1} and C_{a2} are connected in parallel with D_{a1} and D_{a2}, respectively, and absorb the junction capacitors of D_{a1} and D_{a2}. The auxiliary inductor L_a is connected to the midpoint B of the bridge leg, consisting of Q_2 and Q_4. Q_2 and Q_4 are operated in a complementary manner, with 50% duty cycle. Figure 4.3b shows the key waveforms of the auxiliary-current-source network. In each switching period, there are 8 switching modes, whose equivalent circuits are given in Figure 4.4.

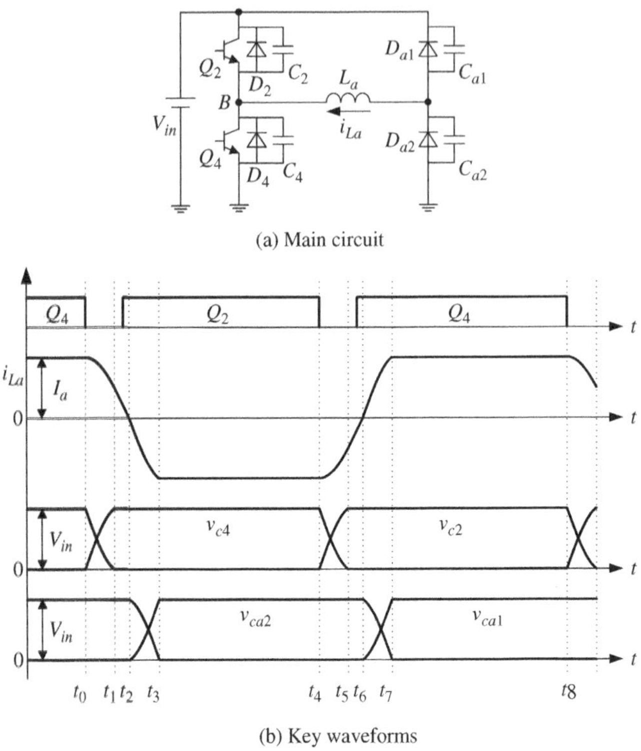

Figure 4.3 Auxiliary-current-source network

Before the analysis, the following assumptions are made: (i) all the switches and diodes are ideal; (ii) all the inductors and capacitors are ideal; and (iii) $C_2 = C_4 = C_r$, $C_{a1} = C_{a2} = C_a$.

Prior to t_0, Q_4 is conducting and the auxiliary inductor current i_{La} is freewheeling through Q_4 and D_{a2}, as shown in Figure 4.4a. The auxiliary inductor current $I_a = \frac{V_{in}}{\sqrt{L_a/(2C_a)}}$, which will be explained later.

At t_0, Q_4 is turned off and i_{La} is transferred from Q_4 to C_2 and C_4, charging C_4 and discharging C_2. Also, L_a resonates with C_2 and C_4, as shown in Figure 4.4b. The auxiliary inductor current and the voltages of auxiliary capacitors are expressed as:

$$v_{c4}(t) = I_a Z_{a1} \sin \omega_{a1}(t - t_0) \tag{4.1}$$

$$v_{c2}(t) = V_{in} - I_a Z_{a1} \sin \omega_{a1}(t - t_0) \tag{4.2}$$

$$i_{La}(t) = I_a \cos \omega_{a1}(t - t_0) \tag{4.3}$$

where $Z_{a1} = \sqrt{L_a/(2C_r)}$ and $\omega_{a1} = 1/\sqrt{2L_a C_r}$. At t_1, the voltage of C_4 increases to the input voltage V_{in} and the voltage of C_2 decays to zero. Thus, the diode D_2 turns on naturally, clamping the voltage of Q_2 at zero.

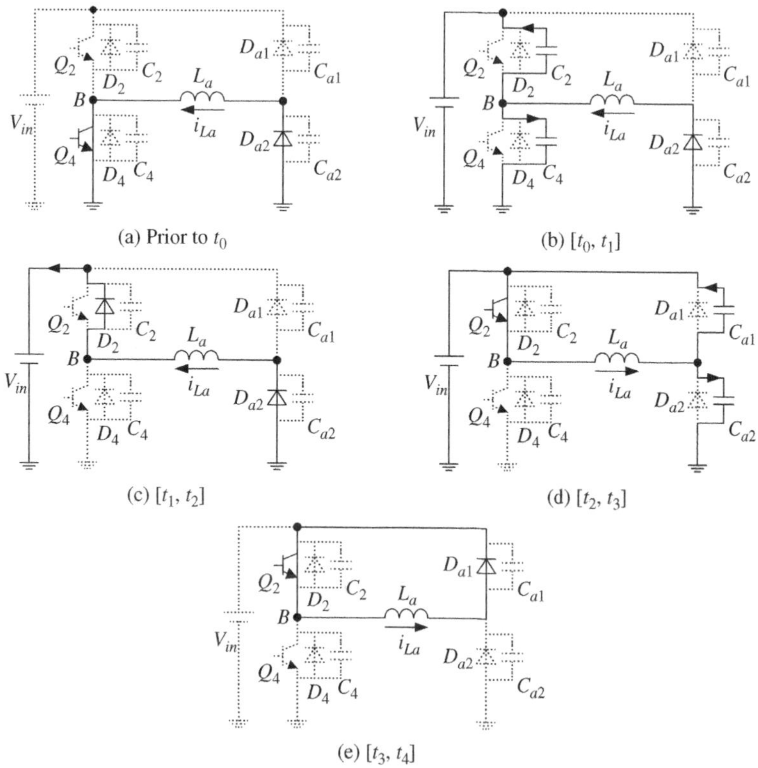

Figure 4.4 Equivalent circuits of the switching modes of an auxiliary-current-source network

When D_2 conducts at t_1, Q_2 can be turned on with zero voltage. The auxiliary inductor current i_{La} flows through D_2 and D_{a2}, as shown in Figure 4.4c. In this mode, the voltage applied on the auxiliary inductor is $-V_{in}$, which forces the auxiliary inductor current to decay linearly; that is:

$$i_{La}(t) = I_{La}(t_1) - \frac{V_{in}}{L_a}(t - t_1) \tag{4.4}$$

At t_2, i_{La} decays to zero. After that time, L_a resonates with C_{a1} and C_{a2}. Moreover, i_{La} increases in the negative direction and C_{a2} is charged and C_{a1} discharged, as shown in Figure 4.4d. The auxiliary inductor current and the voltages of the auxiliary capacitors are expressed as:

$$i_{La}(t) = -\frac{V_{in}}{Z_{a2}} \sin \omega_{a2}(t - t_2) \tag{4.5}$$

$$v_{ca1}(t) = V_{in} \cos \omega_{a2}(t - t_2) \tag{4.6}$$

$$v_{ca2}(t) = V_{in}[1 - \cos \omega_{a2}(t - t_2)] \tag{4.7}$$

where $Z_{a2} = \sqrt{L_a/(2C_a)}$ and $\omega_{a2} = 1/\sqrt{2L_aC_a}$.

At t_3, the voltage of C_{a2} is charged to the input voltage V_{in}, and meanwhile C_{a1} is completely discharged, so D_{a1} conducts naturally. The time duration of t_{23} and the auxiliary inductor current at t_3 are:

$$t_{23} = \frac{\pi}{2}\sqrt{2L_aC_a} \tag{4.8}$$

$$I_{La}(t_3) = -\frac{V_{in}}{Z_{a2}} = -I_a \tag{4.9}$$

After t_3, Q_2 and D_{a1} conduct and the voltage applied on the auxiliary inductor L_a is zero. Thus i_{La} is freewheeling and its values is $-I_a$, as shown in Figure 4.4e.

At t_4, Q_2 is turned off, starting the second half-switching cycle $[t_4, t_8]$, which is similar to the first $[t_0, t_4]$.

From this analysis, it can be concluded that:

1. The maximum value of the auxiliary inductor current, I_a, is determined by the input voltage and the characteristic impedance; that is, $I_a = \frac{V_{in}}{\sqrt{L_a/(2C_a)}}$.
2. The maximum voltage across the auxiliary capacitors is V_{in}.
3. The voltage stress of the auxiliary diodes is V_{in}, while the current stress is I_a.
4. When Q_4 (or Q_2) is turned off, the auxiliary inductor current flows into/out of point B with its maximum value of I_a.

4.3 Operating Principle of a ZVS PWM Full-Bridge Converter with Auxiliary-Current-Source Network

Connecting the auxiliary-current-source network shown in Figure 4.3a to the basic ZVS PWM full-bridge converter shown in Figure 4.1a and sharing the lagging leg, consisting of Q_2 and Q_4, leads to the ZVS PWM full-bridge converter with auxiliary-current-source network shown in Figure 4.5a. With such an arrangement, when Q_4 is turned off the primary current i_p and the auxiliary inductor current i_{La} flow into point B simultaneously, whereas when Q_2 is turned off, i_p and i_{La} flow out of point B simultaneously. Therefore, when either Q_4 or Q_2 is turned off, i_p and i_{La} charge/discharge the parallel capacitors and the lagging leg can realize ZVS in a wide load range.

The popular phase-shift control is adopted here. Q_1 and Q_3 form the leading leg, while Q_2 and Q_4 form the lagging leg. The resonant inductor L_r uses a saturable inductor instead of a linear one. During the switching transition of Q_2 and Q_4, the current of L_r is very small, and the saturable inductor works in a linear manner and has a very large inductance. This large inductance limits the rate of change of the primary current. Upon completion of the switching transition, the saturable inductor enters into a saturable state and the primary current steps up/down to the output inductor current reflected to the primary side. The duty cycle loss is thus reduced nearly to zero and the effective duty cycle becomes equal to the primary duty cycle.

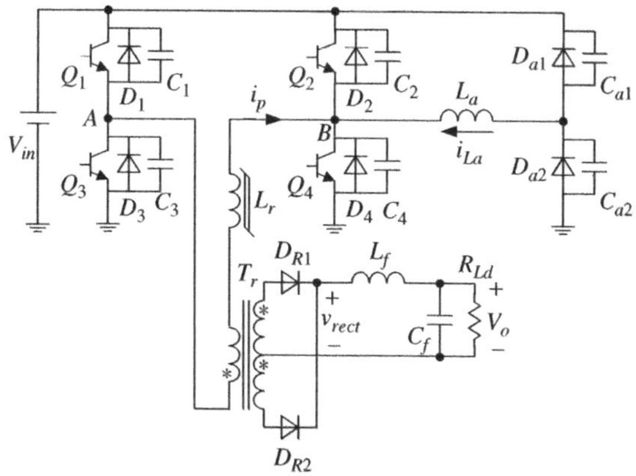

(a) Main circuit

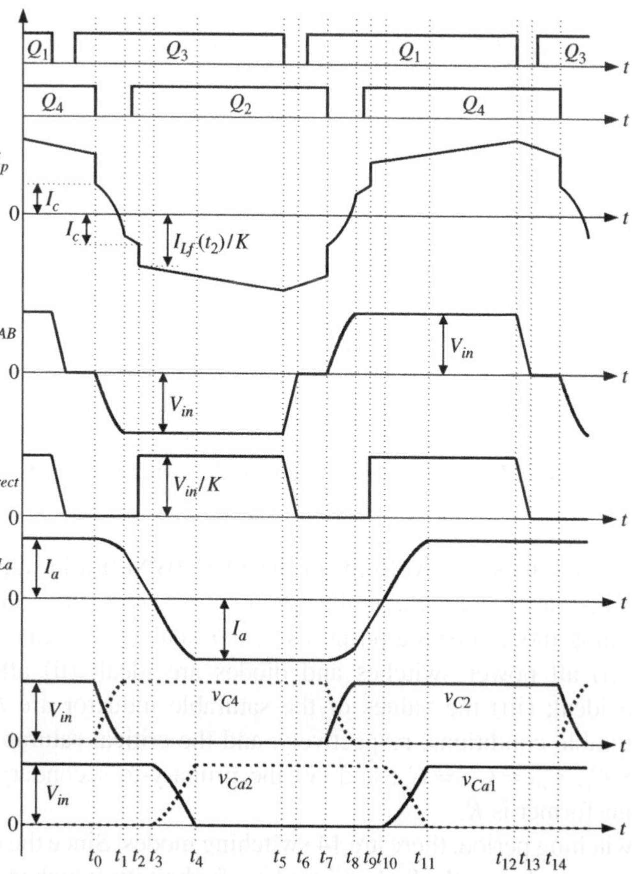

(b) Key waveforms

Figure 4.5 ZVS PWM full-bridge converter with auxiliary-current-source network

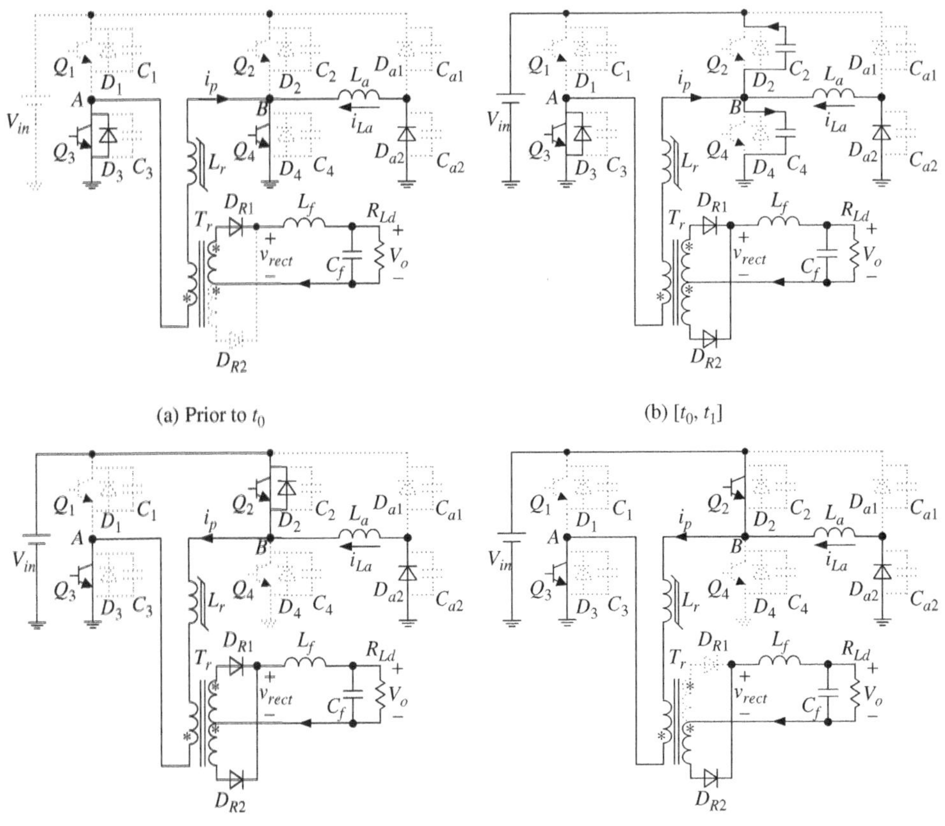

(a) Prior to t_0 (b) $[t_0, t_1]$

(c) $[t_1, t_2]$ (d) $[t_2, t_3]$

Figure 4.6 Equivalent circuits of the switching modes

Figure 4.5b shows the key waveforms of the ZVS PWM full-bridge converter with auxiliary-current-source network, while Figure 4.6 shows the equivalent circuits for each switching mode. For the purposes of our analysis, we make the following assumptions: (i) all power switches and diodes are ideal; (ii) all inductors and capacitors are ideal; (iii) the values of the saturable inductor are L_r and 0 under linear and saturable conditions, respectively, and the critical saturable current is I_c; (iv) $C_2 = C_4 = C_r$, $C_{a1} = C_{a2} = C_a$; and (v) the primary-to-secondary-winding-turns ratio of the transformer is K.

In a given switching period, there are 14 switching modes. Since the first half-period is similar to the second, only the first is given here for brevity, which includes 7 switching modes and 1 initial switching mode:

1. **Mode 0, t_0 (Figure 4.6a):** Prior to t_0, D_3 and Q_4 are conducting, $v_{AB} = 0$, and the primary current i_p is freewheeling. The auxiliary inductor current i_{La} is also

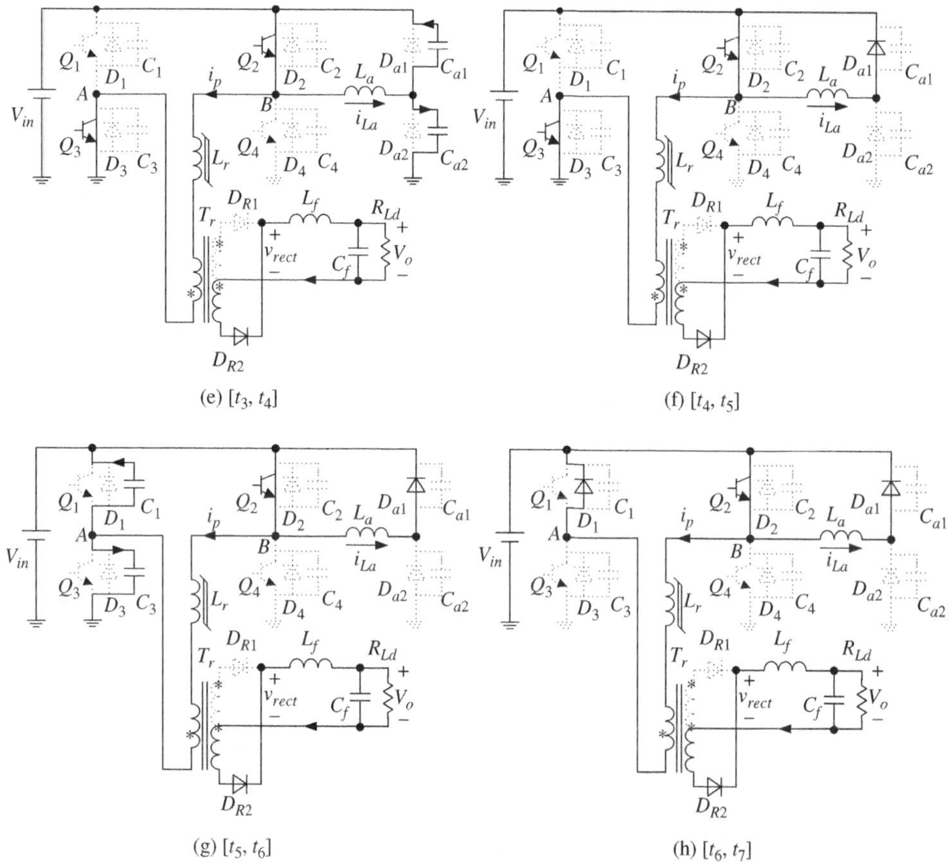

Figure 4.6 (*Continued*)

freewheeling, and flows through Q_4 and D_{a2}. At t_0, $I_{La}(t_0)=I_a$, $V_{C4}(t_0)=0$, $V_{C2}(t_0)=V_{in}$, $V_{Ca1}(t_0)=V_{in}$, $V_{Ca2}(t_0)=0$.

2. **Mode 1, [t_0, t_1] (Figure 4.6b):** At t_0, Q_4 is turned off. i_{La} and i_p charge C_4 and discharge C_2, and $v_{AB}=-v_{C4}$. L_r gets out of saturable condition and enters into linear condition, and i_p jumps down to the critical saturable current I_c and continues decaying. Since i_p is smaller than the reflected output filter inductor current and is not sufficient to power the load, both of the rectifier diodes conduct, making both the primary and the secondary transformer voltage zero. In this mode, the expressions of capacitor voltages and inductor currents are given by:

$$v_{C4}(t) = Z_1(I_c + I_a)\sin\omega_1(t - t_0) \tag{4.10}$$

$$v_{C2}(t) = V_{in} - Z_1(I_c + I_a)\sin\omega_1(t - t_0) \tag{4.11}$$

$$i_p(t) = \frac{L_e}{L_r}(I_c + I_a)[\cos\omega_1(t - t_0) - 1] + I_c \tag{4.12}$$

$$i_{La}(t) = \frac{L_e}{L_a}(I_c + I_a)[\cos\omega_1(t-t_0) - 1] + I_a \tag{4.13}$$

where $L_e = \dfrac{L_r L_a}{L_r + L_a}$, $Z_1 = \sqrt{L_e/(2C_r)}$, $\omega_1 = 1/\sqrt{2L_eC_r}$.

At t_1, the voltage of C_4 increases to V_{in}, the voltage of C_2 reduces to zero, and D_2 conducts naturally. The time interval of mode 1 is:

$$t_{01} = \frac{1}{\omega_1}\sin^{-1}\frac{V_{in}}{Z_1(I_c + I_a)} \tag{4.14}$$

and the currents of L_a and L_r at t_1 are:

$$I_{La}(t_1) = \frac{L_e}{L_a}\sqrt{(I_c + I_a)^2 - \left(\frac{V_{in}}{Z_1}\right)^2} - \frac{L_e}{L_a}(I_c + I_a) + I_a \tag{4.15}$$

$$I_p(t_1) = \frac{L_e}{L_r}\sqrt{(I_c + I_a)^2 - \left(\frac{V_{in}}{Z_1}\right)^2} - \frac{L_e}{L_r}(I_c + I_a) + I_c \tag{4.16}$$

Note that in mode 1, i_p decays and may cross zero and flow in reverse direction. Therefore, D_3 turns off naturally when i_p crosses zero, after which time i_p flows through Q_3.

3. **Mode 2, [t_1, t_2] (Figure 4.6c):** Since D_2 conducts, Q_2 can be turned on with zero voltage. In this mode, $v_{AB} = -V_{in}$ and the voltages across L_a and L_r are both $-V_{in}$, forcing the primary current i_p and auxiliary inductor current i_{La} to decay linearly; that is:

$$i_p(t) = I_p(t_1) - \frac{V_{in}}{L_r}(t - t_1) \tag{4.17}$$

$$i_{La}(t) = I_{La}(t_1) - \frac{V_{in}}{L_a}(t - t_1) \tag{4.18}$$

At t_2, the primary current i_p decays to $-I_c$ and L_r saturates. Thus, the primary current i_p jumps down to the output filter inductor current reflected to the primary side, which equals $-I_{Lf}(t_2)/K$, the rectifier diode D_{R1} turns off, and D_{R2} carries all of the load current.

The time interval of mode 2, t_{12}, is:

$$t_{12} = \frac{L_r}{V_{in}}\left(\frac{I_{Lf}(t_2)}{K} + I_c\right) \tag{4.19}$$

The auxiliary inductor current at t_2 is:

$$I_{La}(t_2) = I_{La}(t_1) - \frac{L_r}{L_a}\left(\frac{I_{Lf}(t_2)}{K} + I_c\right) \tag{4.20}$$

4. **Mode 3, $[t_2, t_3]$ (Figure 4.6d):** In this mode, Q_3 and Q_2 conduct and $v_{AB} = -V_{in}$. The primary side powers the load and the auxiliary inductor current continues decaying, expressed as:

$$i_{La}(t) = I_{La}(t_2) - \frac{V_{in}}{L_a}(t - t_2) \tag{4.21}$$

At t_3, the auxiliary inductor current i_{La} decays to zero and mode 3 ends. The time interval of mode 3 is:

$$t_{23} = L_a I_{La}(t_2)/V_{in} \tag{4.22}$$

5. **Mode 4, $[t_3, t_4]$ (Figure 4.6e):** From t_3, L_a resonates with C_{a1} and C_{a2}. The auxiliary inductor current and two auxiliary capacitor voltages are:

$$i_{La}(t) = -\frac{V_{in}}{Z_2}\sin\omega_2(t - t_3) \tag{4.23}$$

$$v_{Ca2}(t) = V_{in}[1 - \cos\omega_2(t - t_3)] \tag{4.24}$$

$$v_{Ca1}(t) = V_{in}\cos\omega_2(t - t_3) \tag{4.25}$$

where $Z_2 = \sqrt{L_a/(2C_a)}$, $\omega_2 = 1/\sqrt{2L_aC_a}$.

At t_4, the voltage of C_{a2} increases to V_{in}, while the voltage of C_{a1} decays to zero, so D_{a1} begins to conduct naturally. The auxiliary inductor current is:

$$I_{La}(t_4) = -V_{in}/Z_2 = -I_a \tag{4.26}$$

The time interval of mode 4 is:

$$t_{34} = \frac{\pi}{2}\sqrt{2L_aC_a} \tag{4.27}$$

During mode 4, Q_2 and Q_3 continue conducting, $v_{AB} = -V_{in}$, and the primary side powers the load, which is independent of the auxiliary-current-source network.

6. **Mode 5, $[t_4, t_5]$ (Figure 4.6f):** At t_4, D_{a1} conducts, clamping the voltage across the auxiliary inductor L_a at zero, so i_{La} is freewheeling through Q_2 and D_{a1} and its value is kept at $-I_a$.
7. **Mode 6, $[t_5, t_6]$ (Figure 4.6g):** At t_5, Q_3 is turned off with zero voltage thanks to C_1 and C_3. i_p charges C_3 and discharges C_1. The voltages of C_3 and C_1 increase and decay linearly, respectively, because the output filter inductor current is reflected to the primary side and is nearly constant during this interval. At t_6, C_1 is completely discharged, and consequently D_1 conducts naturally. In this mode, i_{La} keeps freewheeling through Q_2 and D_{a1}.
8. **Mode 7, $[t_6, t_7]$ (Figure 4.6h):** When D_1 conducts, Q_1 can be zero-voltage turned on. Since Q_1 and Q_2 are conducting, $v_{AB} = 0$ and i_p is freewheeling through D_1 and Q_2. Meanwhile, i_{La} continues freewheeling through Q_2 and D_{a1}. This mode provides the conditions for the turn-off of Q_2 at t_7, which are similar to those for the turn-off of Q_4: $V_{c2}(t_7) = 0$, $V_{c4}(t_7) = V_{in}$, $I_{La}(t_7) = -I_a$, $V_{ca1}(t_7) = 0$, and $V_{ca2}(t_7) = V_{in}$.

At t_7, Q_2 is turned off, starting the second half-cycle $[t_7, t_{14}]$, which is similar to the first $[t_0, t_7]$.

From this analysis, it can be seen that:

1. During the switching transition of the lagging leg, the auxiliary inductor current i_{La} flows into/out of point B with its maximum-value I_a, which helps the resonant inductor achieve ZVS for the lagging leg.
2. The auxiliary capacitors and auxiliary diodes do not participate in the switching transition of the lagging leg, contributing only to building the maximum-value I_a of the auxiliary inductor. The switching transition of the lagging leg is thus very simple, making it easy to design its parameters. Details will be discussed in Section 4.5.

4.4 Conditions for Achieving ZVS in the Lagging Leg

From mode 1 (see Section 4.3), it can be seen that in order to achieve ZVS for the lagging leg, the following three conditions should be met:

$$V_{C4}(t_1) = Z_1(I_c + I_a)\sin\omega_1 t_{01} = V_{in} \tag{4.28}$$

$$I_p(t_1) = \frac{L_e}{L_r}(I_c + I_a)(\cos\omega_1 t_{01} - 1) + I_c \geq -I_c \tag{4.29}$$

$$I_{La}(t_1) = \frac{L_e}{L_a}(I_c + I_a)(\cos\omega_1 t_{01} - 1) + I_a \geq 0 \tag{4.30}$$

where $t_{01} = t_1 - t_0$.

The condition expressed in Equation 4.28 means that when mode 1 ends, the voltage of C_4 can be increased to V_{in}, allowing D_2 to conduct and thus proving the zero-voltage turn-on condition for Q_2. The condition expressed in Equation 4.29 ensures that the resonant inductor is in linear state, without being saturated at the end of mode 1; otherwise, the primary current would jump to the reflected output filter inductor current and Q_2 would lose the zero-voltage turn-on condition. However, the resonant inductor must be very close to saturation, in order to reduce the time taken to change from linear to saturation state and thus reduce the duty cycle loss. The condition expressed in Equation 4.30 ensures that the auxiliary inductor current does not change its direction at the end of mode 1; otherwise, mode 1 would not be effective.

In practice, $\sin\omega_1 t_{01}$ is kept in the range 0.9–1.0, so as to reduce the value of $Z_1(I_c + I_a)$.

4.5 Parameter Design

This section discusses the design of the parameters related to ZVS achievement of the lagging leg. These include the auxiliary inductor L_a, the auxiliary capacitor C_a, the inductance of the resonant inductor in linear state L_r, the critical saturable current I_c, and the paralleled capacitance C_r of the lagging-leg switches. The preset conditions are: the input voltage V_{in}, the time t_{01} it takes for the parallel capacitor voltage of the outgoing lagging-leg switch to increase from 0 to V_{in}, and the maximum value of the auxiliary inductor current I_a.

4.5.1 *Parameter Selection for the Auxiliary-Current-Source Network*

In order to reduce the conduction loss, the maximum value of the auxiliary inductor current I_a is usually designed to be 10–15% of the load current reflected to the primary side. The characteristic impedance of the auxiliary-current-source network is:

$$Z_2 = \sqrt{\frac{L_a}{2C_a}} = \frac{V_{in}}{I_a} \tag{4.31}$$

The resonant period of the auxiliary-current-source network should be limited in order to ensure that the auxiliary inductor current reaches its maximum value before the turn-off of the lagging-leg switches. As illustrated in Section 4.3, the time it takes for the auxiliary inductor current to increase from 0 to I_a is $\frac{\pi}{2}\sqrt{2L_aC_a}$. Letting the time be $1/N$ of the half-switching period, this gives:

$$\frac{\pi}{2}\sqrt{2L_aC_a} = \frac{T_s}{2N} \tag{4.32}$$

where N is a real number and is greater than one.

According to Equations 4.31 and 4.32, the values of L_a and C_a can be determined as:

$$L_a = \frac{V_{in}T_s}{I_aN\pi} \tag{4.33}$$

$$C_a = \frac{I_a}{V_{in}}\frac{T_s}{2N\pi} \tag{4.34}$$

4.5.2 *Determination of L_r, C_r, and I_c*

Defining:

$$A_g = \sin\omega_1 t_{01} \tag{4.35}$$

and substituting this into Equation 4.28 leads to:

$$Z_1(I_c + I_a)A_g = V_{in} \tag{4.36}$$

It will be seen later that I_c is far smaller than I_a (i.e., $I_c \ll I_a$), so Equation 4.36 can be simplified to:

$$Z_1 I_a A_g = V_{in} \tag{4.37}$$

Substitution of $Z_1 = \sqrt{L_e/(2C_r)}$ and $I_a = V_{in}/\sqrt{L_a/(2C_a)}$ into Equation 4.37 yields:

$$\frac{L_e}{C_r} = \frac{1}{A_g^2}\frac{L_a}{C_a} \tag{4.38}$$

Equation 4.35 can be rewritten as:

$$\omega_1 t_{01} = \sin^{-1} A_g \tag{4.39}$$

Substituting $\omega_1 = 1/\sqrt{2L_e C_r}$ into Equation 4.39 gives:

$$L_e C_r = \frac{1}{2}\left(\frac{t_{01}}{\sin^{-1} A_g}\right)^2 \tag{4.40}$$

According to Equations 4.38 and 4.40, the expressions of L_e and C_r are derived as:

$$L_e = \frac{1}{\sqrt{2}} \frac{t_{01}}{A_g \sin^{-1} A_g} \sqrt{\frac{L_a}{C_a}} \tag{4.41}$$

$$C_r = \frac{1}{\sqrt{2}} \frac{A_g t_{01}}{\sin^{-1} A_g} \sqrt{\frac{C_a}{L_a}} \tag{4.42}$$

Then, according to $L_e = \dfrac{L_a L_r}{L_a + L_r}$, the expression of L_r is derived as:

$$L_r = \frac{L_a L_e}{L_a - L_e} \tag{4.43}$$

Similarly, given $I_c \ll I_a$ and according to Equation 4.29, we get:

$$I_c \geq \frac{L_e}{2L_r} I_a (1 - \cos \omega_1 t_{01}) = \frac{L_e}{2L_r} \frac{V_{in}}{\sqrt{\dfrac{L_a}{2C_a}}} \left(1 - \sqrt{1 - A_g^2}\right) \tag{4.44}$$

4.5.3 *Design Example*

The main specifications of the prototype designed in this chapter are as follows: input voltage $V_{in} = 537\,\text{V}^{+15\%}_{-20\%}$, which is rectified and filtered from a three-phase 380 V/ 50 Hz ac line voltage; output voltage $V_o = 52.8$ V; rated output current $I_o = 50$ A; transformer primary-to-secondary-winding-turns ratio $K = 5.5$; power switches are insulated gate bipolar transistors (IGBTs) with fall times $t_f = 0.7\,\mu$s – here we choose $t_{01} = 2.5 \cdot t_f = 1.75\,\mu$s for the purpose of reducing the turn-off loss resulting from the current tail; switching frequency $f_s = 30$ kHz.

The design procedure is:

1. At the nominal input voltage $V_{in} = 537$ V, I_a is set to 12.5% of the output current reflected to the primary side; I_a is then 12.5% × 50/5.5 = 1.136 A.
2. Letting $N = 5$ and substituting it and $I_a = 1.136$ A into Equations 4.33 and 4.34 leads to $L_a = 1.003$ mH and $C_a = 2.245$ nF. We choose $L_a = 910\,\mu$H and $C_a = 2.2$ nF, so $I_a = 1.18$ A at the nominal input voltage $V_{in} = 537$ V.

3. Letting $A_g = 0.9$ and substituting it and the obtained values of L_a and C_a into Equations 4.41–4.44 leads to $L_r = 4.808$ mH, $I_c = 0.055$ A, and $C_r = 1.475$ nF. We choose $L_r = 5$ mH, $I_c = 0.07$ A, and $C_r = 1.5$ nF. It is obvious that $I_c \ll I_a$, which illustrates that the assumption in Section 4.5.2 is reasonable.
4. Substituting these parameters into the left side of Equation 4.30, we have $I_{La}(t_1) = 0.684$ A. Obviously, $I_{La}(t_1)$ is greater than zero, and the condition of Equation 4.30 is satisfied.

From this analysis, it can be seen that when the load current is greater than KI_c (i.e., the reflected load current is greater than the critical saturation current of the resonant inductor), the lagging leg can realize ZVS. Here $I_c = 0.07$ A, so the minimum load current required to achieve ZVS for the lagging leg is $5.5 \times 0.07 = 0.385$ A, which represents 0.77% of the full load. The proposed ZVS PWM full-bridge converter with auxiliary-current-source network can thus achieve ZVS for the lagging leg over nearly the entire load range.

4.6 Secondary Duty Cycle Loss and Selection of Dead Time for the Drive Signals of the Lagging Leg

This section discusses duty cycle loss and selection of the dead time between the drive signals of lagging-leg switches according to the parameters designed in Section 4.5.3.

4.6.1 Secondary Duty Cycle Loss

Secondary duty cycle loss has two parts. The first corresponds to the time duration of mode 1, t_{01}, in which the voltage of C_2 or C_4 increases from zero to V_{in}. During t_{01}, the primary side does not provide any load current. The second part corresponds to the time duration of mode 2, t_{12}, in which D_4 or D_2 conducts and the resonant inductor has not saturated. The primary side still does not provide load current. As a consequence, v_{AB} during t_{01} and t_{12} does not reflect to the secondary side, resulting in duty cycle loss. The duty cycle losses corresponding to t_{01} and t_{12} are defined as D_{loss1} and D_{loss2}, expressed as $D_{loss1} = \frac{t_{01}}{T_s/2}$ and $D_{loss2} = \frac{t_{12}}{T_s/2}$, respectively.

D_{loss1} is unavoidable because t_{01} is the time it takes to achieve ZVS. We shall look at D_{loss2} here. Substituting the selected parameters into Equations 4.14 and 4.19, we obtain the curves of t_{01} and t_{12} over the input voltage range, as well as the curve of t_{02}, the sum of t_{01} and t_{12} (Figure 4.7). As can be seen, as the input voltage increases from 430 to 618 V, t_{01} increases from 1.504 to 1.555 µs – a very small variation in t_{01}; meanwhile, t_{12} reduces from 0.808 to 0.279 µs, and the corresponding D_{loss2} is 0.048 and 0.017, respectively. Obviously, D_{loss2} is very small. Noted that the calculated t_{01} is slightly smaller than the preset 1.75 µs; this is because I_c is neglected in the parameter design.

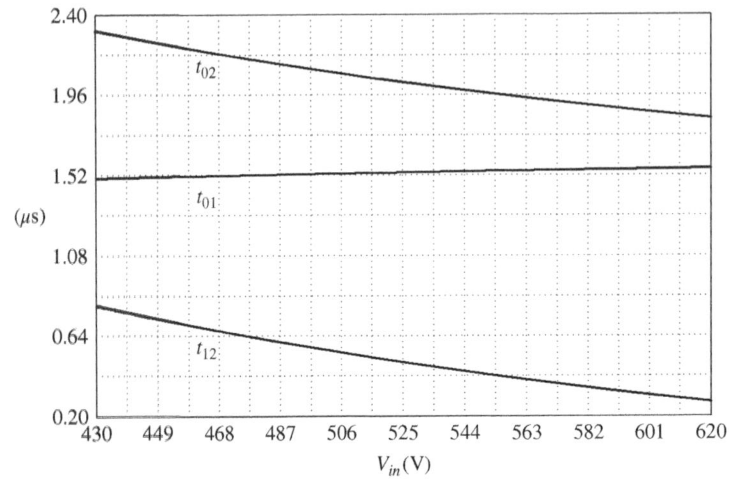

Figure 4.7 Curves of t_{01}, t_{12}, and t_{02} over the input voltage range

4.6.2 Selection of Dead Time between Drive Signals of the Lagging Leg

In order to achieve zero-voltage turn-on for the lagging-leg switches, the voltage across the paralleled capacitor of the incoming switch should be completely discharged prior to its turn-on. This means that a dead time t_d should be introduced between the turn-off of the outgoing switch and the turn-on of the incoming switch. This dead time must meet the following requirement:

$$t_{01} \leq t_d \leq t_{01} + t_{12} = t_{02} \tag{4.45}$$

As can be seen in Figure 4.7, t_{02} decays with increasing of input voltage. In order to achieve zero-voltage turn-on for the lagging-leg switches, t_d can thus be slightly smaller than the minimum t_{02} and larger than the maximum t_{01}, which both correspond to the maximum input voltage.

4.6.3 Comparison with Full-Bridge Converter with Saturable Inductor

Reference [2] proposes using a saturable inductor instead of a linear one as the resonant inductor, as shown in Figure 4.8. When the lagging-leg switch Q_4 (or Q_2) turns off, we get the following expressions:

$$v_{C4}(t) = \sqrt{\frac{L_r}{2C_r}} I_c \sin \omega_3 (t - t_0) \tag{4.46}$$

$$i_{Lr}(t) = I_c \cos \omega_3 (t - t_0) \tag{4.47}$$

where $\omega_3 = 1/\sqrt{2L_r C_r}$. At t_1, the voltage of C_4 increases from zero to V_{in}.

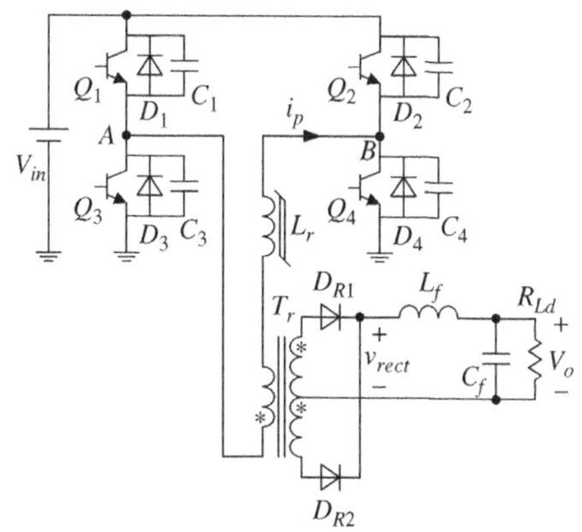

(a) Main circuit

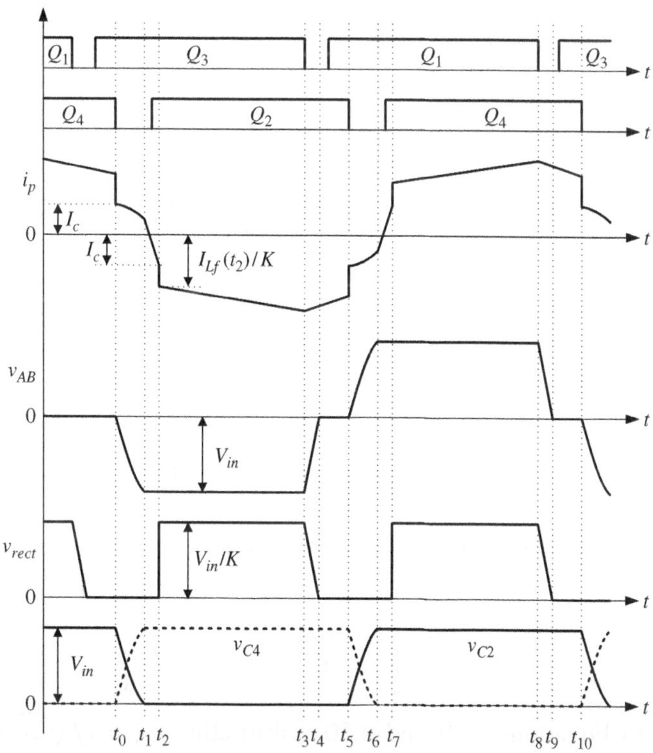

(b) Key Waveforms

Figure 4.8 ZVS PWM full-bridge converter adopting saturable inductor [2]

It can be seen from Equation 4.46 that, in order to achieve ZVS for the lagging leg, the following condition should be satisfied:

$$\sqrt{\frac{L_r}{2C_r}}I_c \geq V_{in} \tag{4.48}$$

Apparently, the higher the input voltage, the larger the energy required to achieve ZVS for the lagging leg.

As can be seen from Equation 4.46, once $\sqrt{L_r/(2C_r)}I_c$ is selected, the higher the input voltage, the longer the time t_{01} required for the voltage of C_4 to increase from zero to V_{in}. The maximum t_{01} is defined as $t_{01\_\max}$. For a fair comparison, at the maximum input voltage $V_{in\max}$ we let:

$$\sin \omega_3 t_{01\_\max} = 0.9 \tag{4.49}$$

and according to Equation 4.46, we have:

$$\sqrt{\frac{L_r}{2C_r}}I_c = \frac{V_{in\max}}{0.9} \tag{4.50}$$

According to Equations 4.46, 4.49, and 4.50, t_{01} can be derived as:

$$t_{01} = t_{01\_\max}\frac{\sin^{-1}\left(\dfrac{0.9V_{in}}{V_{in\max}}\right)}{\sin^{-1}0.9} \tag{4.51}$$

At t_1, the saturable inductor current is:

$$I_{L_r}(t_1) = I_c\sqrt{1-\left(\frac{0.9V_{in}}{V_{in\max}}\right)^2} \tag{4.52}$$

Thus, the time required for the saturable inductor to get out of linear state is:

$$t_{12} = \frac{L_r I_c}{V_{in}}\left[1+\sqrt{1-\left(\frac{0.9V_{in}}{V_{in\max}}\right)^2}\right] \tag{4.53}$$

According to Equations 4.49 and 4.50, substituting $\omega_3 = 1/\sqrt{2L_rC_r}$ into them, we get:

$$L_r I_c = \frac{V_{in\max}t_{01\_\max}}{0.9\cdot\sin^{-1}0.9} \tag{4.54}$$

Substitution of Equation 4.54 into Equation 4.53 yields:

$$t_{12} = \frac{V_{in\max}t_{01\_\max}}{0.9V_{in}\cdot\sin^{-1}0.9}\left[1+\sqrt{1-\left(\frac{0.9V_{in}}{V_{in\max}}\right)^2}\right] \tag{4.55}$$

Letting $t_{01\_max} = 1.75\,\mu s$, we get $t_{01} = 1.356\,\mu s$ at the nominal input voltage $V_{in} = 537$ V. At the highest and lowest input voltages, t_{12} is equal to 1.929 and 3.698 μs, respectively. At the lowest input voltage, t_{12} in the full-bridge converter with a saturable inductor is 4.58 times that in the full-bridge converter with auxiliary-current-source network proposed in this chapter (0.808 μs). That is to say, the duty cycle loss in the proposed converter is only 1/4.58 that in the converter with saturable inductor proposed in reference [2].

4.7 Experimental Verification

In order to demonstrate the operating principle of the ZVS PWM full-bridge converter with auxiliary-current-source network, a 52.8 V/50 A output prototype converter was fabricated and tested in the lab. The main specifications are as follows:

- input dc voltage $V_{in} = 537$ V, with variation range from 430 to 618 V;
- output voltage $V_o = 52.8$ V; and
- output current $I_o = 50$ A.

The key power devices and elements of the prototype are:

- transformer primary-to-secondary-winding-turns ratio $K = 5.5$;
- saturable inductor $L_r = 5.0$ mH in linear state and critical saturable current $I_c = 0.07$ A;
- paralleled capacitors $C_2 = C_4 = C_r = 1.5$ nF;
- auxiliary inductor $L_a = 910\,\mu H$;
- auxiliary capacitors $C_{a1} = C_{a2} = 2.2$ nF;
- output filter inductor $L_f = 30\,\mu H$;
- output filter capacitor $C_f = 10\,000\,\mu F$;
- power switches (IGBT): VII50-12Q3;
- output rectifier diodes: DSEI2 × 61-06C; and
- switching frequency $f_s = 30$ kHz.

Figure 4.9 shows the experimental waveforms at full load, with Figure 4.9a showing the waveforms of the primary voltage v_{AB} and primary current i_p. As can be seen, the resonant inductor works in linear state during the switching transition of the lagging leg, and once the voltage of the paralleled capacitor of the outgoing switch (v_{C4} or v_{C2}) increases to V_{in} it quickly enters into saturable state, which leads to a very small duty cycle loss. This can also be seen from the waveforms of the voltage across the resonant inductor and the primary current, as shown in Figure 4.9b. Figure 4.9c shows the waveforms of the primary voltage v_{AB} and the secondary rectified voltage v_{rect}, which also indicates that the secondary duty cycle is nearly equal to the primary duty cycle, and only the duty cycle loss corresponding to the time it takes for the voltage of the parallel capacitor (v_{C4} or v_{C2}) of the outgoing lagging-leg switch

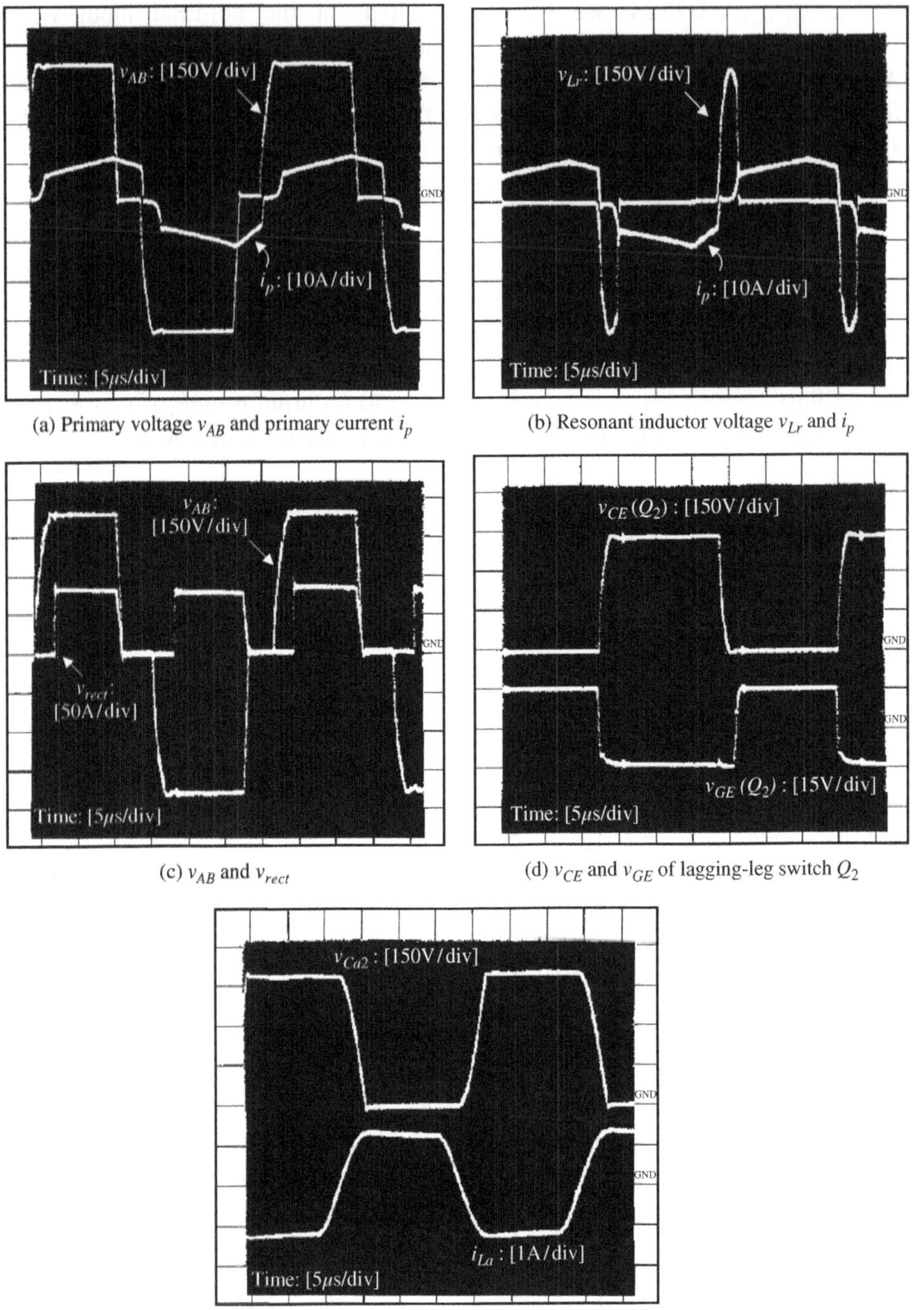

(a) Primary voltage v_{AB} and primary current i_p

(b) Resonant inductor voltage v_{Lr} and i_p

(c) v_{AB} and v_{rect}

(d) v_{CE} and v_{GE} of lagging-leg switch Q_2

(e) Waveforms of the voltage of auxiliary capacitor C_{a2} and auxiliary inductor current

Figure 4.9 Experimental waveforms at full load

to increase from zero to V_{in} occurs; this loss cannot be avoided. Figure 4.9d shows the waveforms of the collector-to-emitter voltage v_{CE} and the gate drive signal v_{GE} of the lagging-leg power switch Q_2. As can be seen, after v_{CE} reduces to zero and the antiparallel diode conducts, Q_2 is triggered on. Q_2 thus realizes zero-voltage turn-on. When Q_2 is turned off, v_{CE} increases slowly, so Q_2 realizes zero-voltage turn-off. The lagging leg thus realizes ZVS. Figure 4.9e shows the auxiliary capacitor voltage and auxiliary inductor current. As can be seen, the auxiliary capacitor voltage stays at V_{in} during the switching transition of the lagging leg, which means that the auxiliary capacitors do not participate in this and only help in building the maximum auxiliary inductor current. Also, the maximum auxiliary inductor current is determined by the auxiliary-current-source network and is independent of the load current. The measured maximum auxiliary inductor current is 1.35 A, which is far lower than the load current reflected to the primary side.

4.8 Other Auxiliary-Current-Source Networks for ZVS PWM Full-Bridge Converters

Based on the current-enhancement principle proposed in Section 4.2, there are various auxiliary-current-source networks that can be used with a ZVS PWM full-bridge converter, which can be cataloged into four types: uncontrolled auxiliary current magnitude, controlled auxiliary current magnitude, auxiliary current magnitude proportional to primary duty cycle, and load-adaptive auxiliary current magnitude. Some auxiliary networks can achieve a resonant inductor current that is adaptive to the load current.

4.8.1 Auxiliary-Current-Source Networks with Uncontrolled Auxiliary Current Magnitude

The auxiliary current magnitude of the auxiliary-current-source network presented in Section 4.2 is uncontrolled. Figure 4.10 shows two others. An auxiliary-current-source network composed of an auxiliary inductor L_a and two auxiliary capacitors C_{a1} and C_{a2} [3] is shown in Figure 4.10a, where C_{a1} and C_{a2} are two divided capacitors with relatively large capacitances and share the input voltage evenly; that is, $V_{ca1} = V_{ca2} = V_{in}/2$. When Q_4 conducts, the voltage across the auxiliary inductor L_a is $V_{in}/2$, forcing the auxiliary inductor current i_{La} to increase linearly; likewise, when Q_2 conducts the voltage across L_a is $-V_{in}/2$, forcing i_{La} to decay linearly. i_{La} is a bipolar triangular waveform with a magnitude of $I_a = V_{in}T_s/(8L_a)$, as shown in Figure 4.11.

Removing the capacitor C_{a2} shown in Figure 4.10a leads to the simplified auxiliary-current-source network shown in Figure 4.10b [4]. Since Q_2 and Q_4 are switched in complementary manner with 50% duty cycle, the average voltage of midpoint B in a switching period is $V_{in}/2$. Meanwhile, the average voltage across the

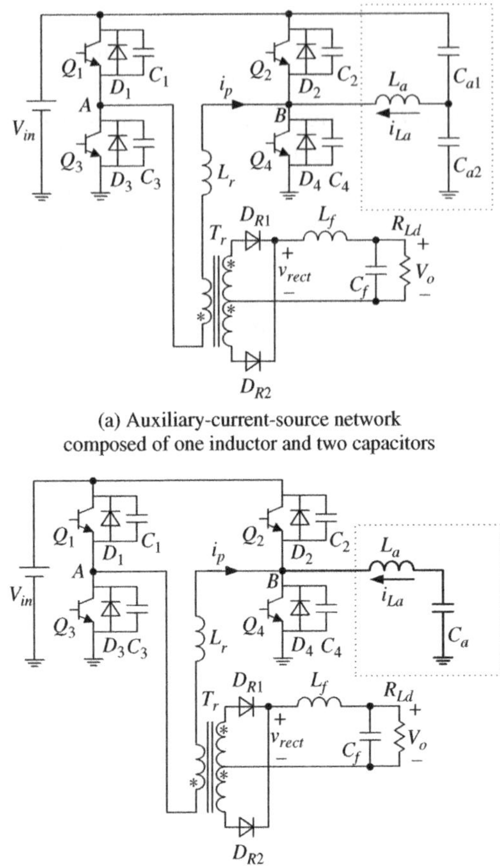

(a) Auxiliary-current-source network
composed of one inductor and two capacitors

(b) Auxiliary-current-source network
composed of one inductor and one capacitor

Figure 4.10 Auxiliary-current-source networks with uncontrolled auxiliary current magnitude

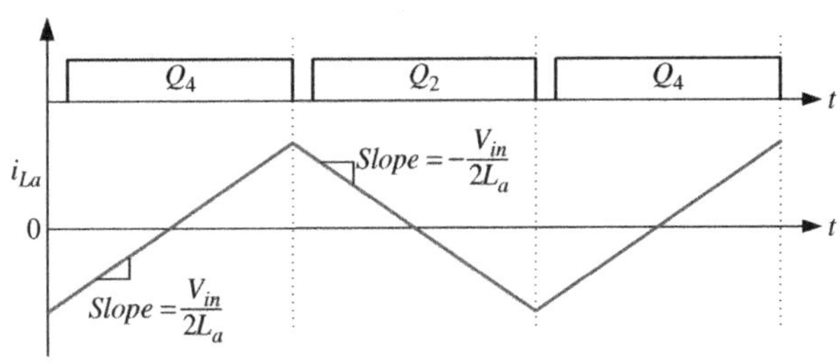

Figure 4.11 Waveforms of auxiliary-current-source networks with uncontrolled auxiliary current magnitude

auxiliary inductor is zero at steady state. Thus the voltage of the auxiliary capacitor C_a (with large capacitance) is $V_{ca} = V_{in}/2$ and the right-terminal voltage of the auxiliary inductor is $V_{in}/2$, which is the same as that in Figure 4.10a. The auxiliary inductor current is therefore also a bipolar triangular waveform, as in Figure 4.11.

4.8.2 *Auxiliary-Current-Source Networks with Controlled Auxiliary Current Magnitude*

For the auxiliary-current-source networks shown in Figures 4.3 and 4.10, when the values of the auxiliary inductor and capacitors are determined, the auxiliary inductor magnitude is also determined, and it is found to be independent of load current. At heavy load, the energy stored in the resonant or leakage inductor is large enough to achieve ZVS for the lagging leg, even without the help of the auxiliary inductor. In this case, the additional conduction loss resulting from the auxiliary inductor current will thus lead to a decrease of conversion efficiency.

In order to regulate the auxiliary inductor current according to load current, the two auxiliary capacitors in Figure 4.10a can be replaced by two switches Q_{a1} and Q_{a2} (including their antiparalleled diodes D_{a1} and D_{a2}), as shown in Figure 4.12a [5]. Figure 4.12b gives the key waveforms. The duty cycle of Q_{a1} and Q_{a2} increases with the decrease in load current. This means that the duty cycle of Q_{a1} and Q_{a2} is large at light load, providing a larger auxiliary inductor current to ensure ZVS for the lagging leg, and is small at heavy load, reducing the auxiliary inductor current and thus the conduction loss.

This auxiliary-current-source network has the following advantages: (i) its work time is very short, so the conduction losses in the auxiliary switches, inductor, and lagging leg are small; and (ii) the currents in the auxiliary switches and inductor are far smaller than the load current and the voltage stress of the auxiliary switches is the input voltage V_{in}. However, it also has some drawbacks, including the fact that: (i) two more switches are required, together with corresponding drive circuits; (ii) the two auxiliary switches are hard turn-off, resulting in turn-off loss; and (iii) a load current sensor is required for control of the two auxiliary switches.

4.8.3 *Auxiliary-Current-Source Network with Auxiliary Current Magnitude Proportional to Primary Duty Cycle*

The simplest auxiliary-current-source network is an auxiliary inductor L_a, which is connected to the two midpoints of the bridge legs, as shown in Figure 4.13a [6]. This auxiliary inductor can be realized by the magnetizing inductor of the transformer, so the circuit is very simple. When the primary voltage v_{AB} is V_{in}, the auxiliary inductor current i_{La} increases linearly; when v_{AB} is $-V_{in}$, i_{La} decays linearly; and when v_{AB} is zero, i_{La} remains unchanged. Figure 4.13b shows the key waveforms, from which it can be seen that the magnitude of i_{La} is proportional to the primary duty cycle.

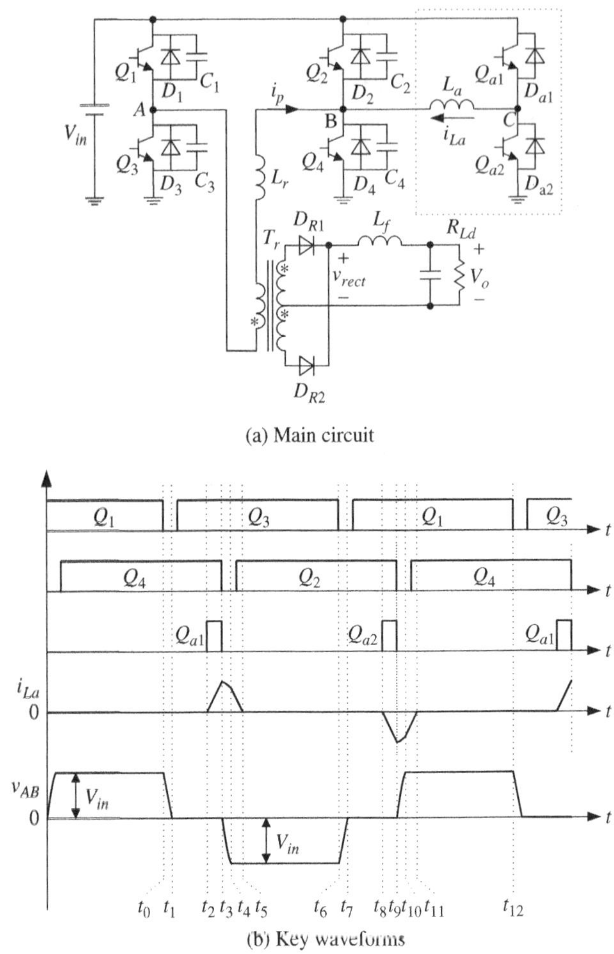

Figure 4.12 Auxiliary-current-source network with controlled auxiliary current magnitude

As analyzed in Chapter 3, the leakage inductor or external resonant inductor results in duty cycle loss, which increases with the load current; as a consequence, the primary duty cycle is increased for the desired output voltage when the load current increases, and thus the magnitude of i_{La} increases. At heavy load, the energy stored in the resonant inductor is enough to achieve ZVS for the lagging leg, even without the help of the auxiliary inductor current. At light load, the energy stored in the resonant inductor is insufficient to achieve ZVS, and a larger auxiliary inductor current is required. However, the magnitude of the auxiliary inductor current is proportional to the load current and in reverse trend with the required magnitude. If the magnitude of the auxiliary inductor current is designed to ensure ZVS at light load, it is very large at heavy load, leading to higher conduction loss, and thus the conversion efficiency at heavy load is reduced.

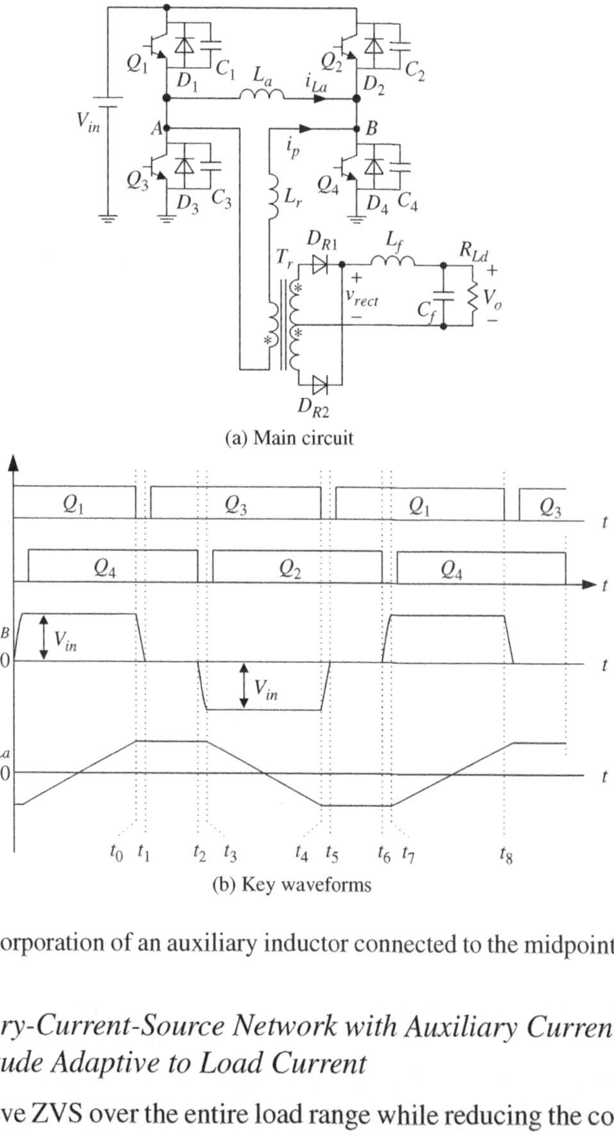

Figure 4.13 Incorporation of an auxiliary inductor connected to the midpoints of bridge legs

4.8.4 Auxiliary-Current-Source Network with Auxiliary Current Magnitude Adaptive to Load Current

In order to achieve ZVS over the entire load range while reducing the conduction loss, the magnitude of the auxiliary inductor current should be adaptive to the load current; that is, small at heavy load and large at light load. This can be realized by letting the duty cycle of the voltage across the auxiliary inductor be complementary to the primary duty cycle.

It is known that the voltage across the midpoints of two bridge legs v_{AB} is equal to $v_A - v_B$. Thus, the voltage at which the duty cycle is complementary to v_{AB} can be obtained from $v_A + v_B$. For the phase-shift-controlled full-bridge converter, the voltages of the midpoints of the bridge leg v_A and v_B are both dc square waves with 180° electrical angle and contain a dc component of $V_{in}/2$. So $v_A + v_B$ has a dc component of V_{in} ($=2 \cdot V_{in}/2$). What we need is a pure ac voltage, so the dc component should be

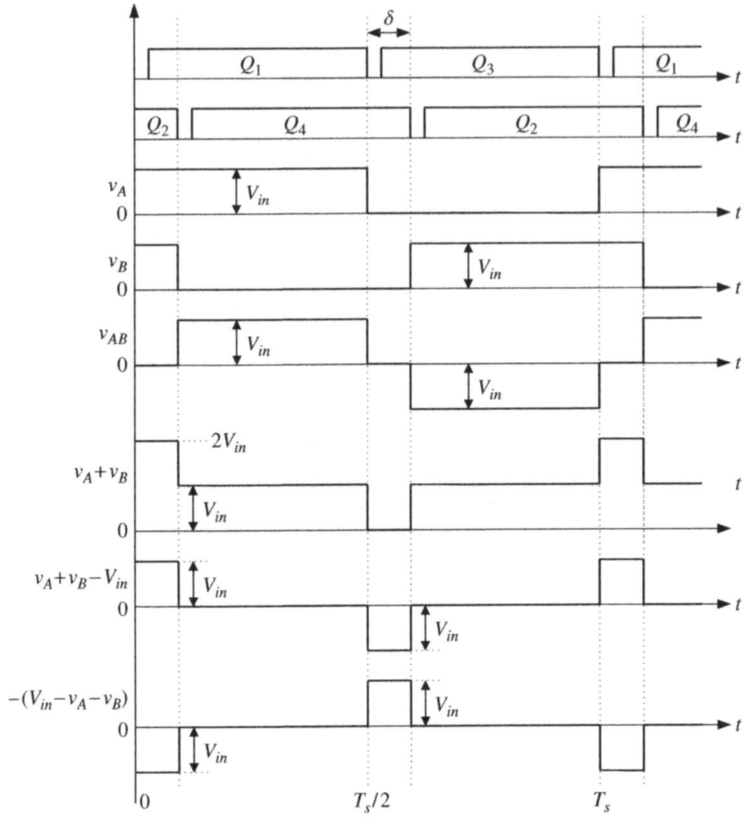

Figure 4.14 Waveforms of voltages with complementary duty cycles

removed. Thus, the desired voltage is $v_A + v_B - V_{in}$. Figure 4.14 shows the corresponding waveforms. In fact, what concerns us is whether the duty cycle of the obtained ac voltage is complementary to that of v_{AB}, rather than its magnitude or polarity. In other words, the required ac voltage can be $(V_{in} - v_A - v_B)/m$, where m is either a positive real number or a negative real number, its value selected according to the specific circuit. Figure 4.14 gives the waveforms of $V_{in} - v_A - v_B$ and $-(V_{in} - v_A - v_B)$.

In the preceding discussion, it is assumed that the transformer primary winding is connected to the midpoint of the two bridge legs (i.e., that the primary voltage is v_{AB}) and that $(V_{in} - v_A - v_B)/m$ will be applied to the auxiliary inductor. In fact, $(V_{in} - v_A - v_B)/m$ can also be applied to the transformer primary winding, when v_{AB} is applied to the auxiliary inductor. By observing the waveforms in Figure 4.14, it can be seen that for the former arrangement the output voltage decays with an increase in phase shift δ. This is the control logic of popular controllers such as UC3875, UC3879, and UC3895. For the latter arrangement, the output voltage increases with an increase in shifted-phase δ. If controllers such as UC3875, UC3879, and UC3895 are adopted, some additional circuits are required to reverse the control logic.

4.8.4.1 Transformer Primary Winding Voltage is v_{AB}

Figure 4.15 shows the main circuit and key waveforms of a ZVS PWM full-bridge converter with auxiliary current source magnitude adaptive to load [7], where T_a is the auxiliary transformer with turns ratio of 1, and where C_{a1} and C_{a2} are the divided capacitors, sharing the input voltage evenly; that is, $V_{ca1} = V_{ca2} = V_{in}/2$. The voltage

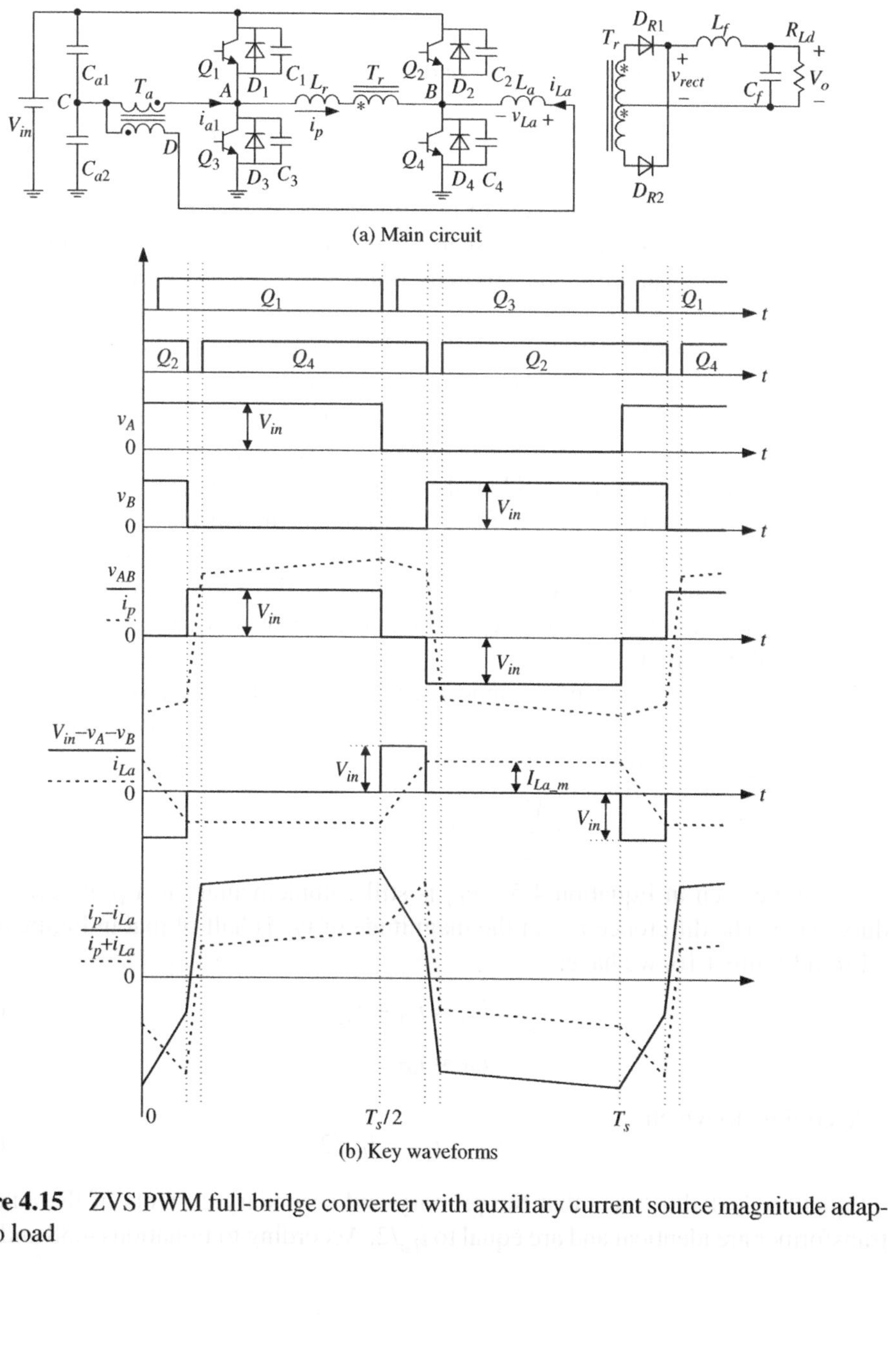

Figure 4.15 ZVS PWM full-bridge converter with auxiliary current source magnitude adaptive to load

applied on the auxiliary inductor is:

$$v_{La} = \frac{V_{in}}{2} - v_{CD} - v_B = \frac{V_{in}}{2} - \left(v_A - \frac{V_{in}}{2}\right) - v_B = V_{in} - v_A - v_B \quad (4.56)$$

Figure 4.15b shows the waveform of $V_{in} - v_A - v_B$, which is complementary to v_{AB} in term of duty cycle.

When v_{AB} equals $+V_{in}$ or $-V_{in}$, the voltage applied on the auxiliary inductor $v_{La} = 0$ and the auxiliary inductor current i_{La} remains unchanged; when $v_{AB} = 0$, v_{La} equals $+V_{in}$ or $-V_{in}$ and i_{La} linearly increases or decays, respectively. The magnitude of i_{La} is:

$$I_{La\_m} = \frac{1}{2}\frac{V_{in}}{L_a}(1 - D_p)\frac{T_s}{2} = \frac{V_{in}}{4L_a f_s}(1 - D_p) \quad (4.57)$$

where D_p is the primary duty cycle of the full-bridge converter.

Referring to Figure 4.15a, it can be seen that the auxiliary transformer secondary current is equal to i_{La}. Therefore, the auxiliary transformer primary current i_{a1} is equal to i_{La}, due to the turns ratio being 1. As can be seen in Figure 4.15b, when the lagging-leg switch Q_4 or Q_2 is turned off, the primary current i_p and auxiliary inductor current i_{La} flow into and out of point B, respectively; when the leading-leg switch Q_3 or Q_1 is turned off, the primary current i_p and auxiliary transformer secondary current i_{a1} flow into and out of point A, respectively. This means that the auxiliary inductor not only helps the lagging leg realize ZVS but also helps the leading leg realize ZVS, and that its magnitude is adaptive to load. The waveforms of the current $i_p - i_{a1}$ ($= i_p - i_{La}$), which flows out of midpoint A, and of the current $i_p + i_{La}$, which flows into midpoint B, are shown in Figure 4.15b. It can be seen that $i_p - i_{La}$ is different from $i_p + i_{La}$. Thus, the current stresses of the leading-leg and lagging-leg switches are different. This is a drawback of this circuit.

The auxiliary inductor can be connected to the midpoint of the two divide capacitors and the interconnected point of the two windings of the auxiliary transformer, as shown in Figure 4.16 [8]. The voltage of the auxiliary inductor v_{La} is:

$$v_{La} = \frac{V_{in}}{2} - v_{DB} - v_B = \frac{V_{in}}{2} - \frac{1}{2}(v_A - v_B) - v_B = \frac{1}{2}(V_{in} - v_A - v_B) \quad (4.58)$$

As can be seen in Equation 4.58, v_{La} is still complementary to v_{AB} in term of the duty cycle. The difference is that the magnitude of v_{La} is half of that in Figure 4.15a.

From Figure 4.16, we have:

$$i_{La} = i_{a1} + i_{a2} \quad (4.59)$$

$$i_{a1} = i_{a2} \quad (4.60)$$

According to which:

$$i_{a1} = i_{a2} = i_{La}/2 \quad (4.61)$$

Equation 4.61 illustrates that the primary and secondary currents of the auxiliary transformer are identical and are equal to $i_{La}/2$. According to Equations 4.58 and 4.61,

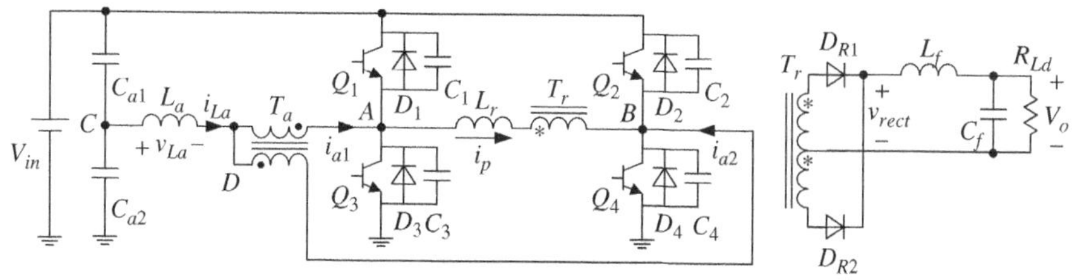

Figure 4.16 ZVS PWM full-bridge converter with auxiliary current source magnitude adaptive to load, proposed in reference [8]

in order to obtain the same auxiliary current magnitude, the auxiliary inductance in Figure 4.16 should be one-quarter that in Figure 4.15a.

In Figure 4.16, the divided capacitor C_{a1} can be removed and the voltage of the remaining capacitor C_{a2} will still be $V_{in}/2$. The simplified converter is shown in Figure 4.17.

The function of capacitor C_a in Figure 4.17 is to provide the voltage of $V_{in}/2$. This can be realized by two dc blocking capacitors, C_{b1} and C_{b2}, as shown in Figure 4.18 [9]. Both of the dc blocking capacitors have voltage $V_{in}/2$, and their polarities are marked in Figure 4.18.

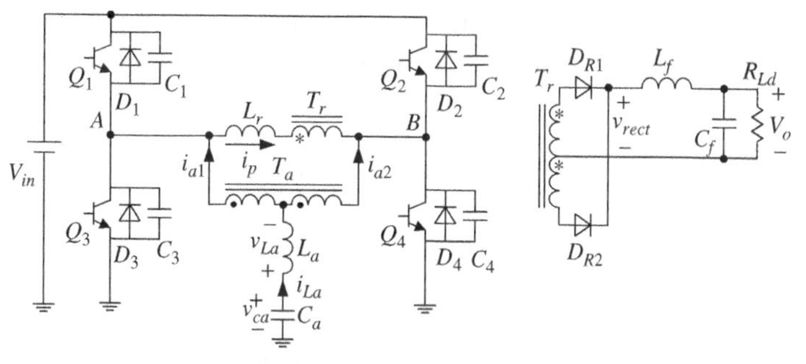

Figure 4.17 Simplified topology of the full-bridge converter proposed in reference [8]

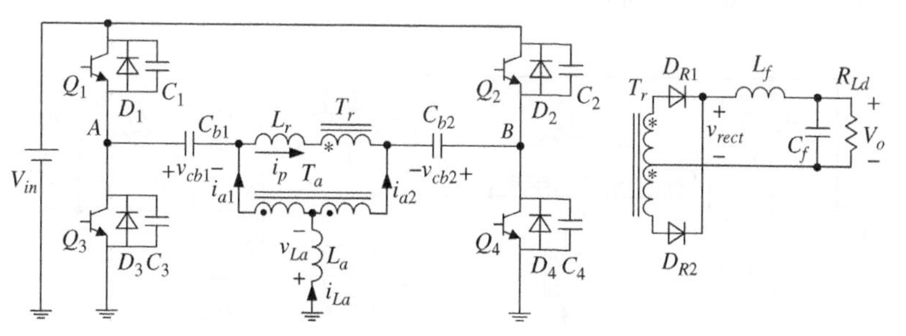

Figure 4.18 ZVS PWM full-bridge converter with auxiliary current source magnitude adaptive to load using two dc blocking capacitors

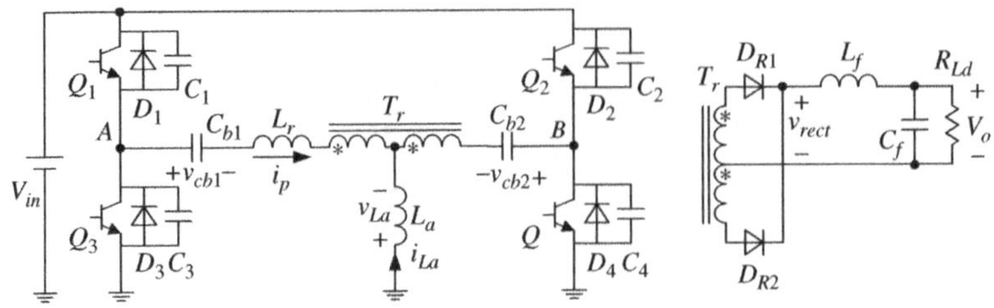

Figure 4.19 ZVS PWM full-bridge converter with auxiliary current source magnitude adaptive to load using the main transformer

By observing Figure 4.18, it can be seen that if the resonant inductor L_r is small enough to be neglected, the function of the auxiliary transformer T_a can be realized by the main transformer T_r, by splitting the primary winding of the main transformer into two windings with the same turns, as shown in Figure 4.19. This is the full-bridge converter proposed in reference [10].

4.8.4.2 Transformer Primary Winding Voltage is $(V_{in} - v_A - v_B)/m$

Figure 4.20 shows the full-bridge converter proposed in reference [11], in which the coupled inductor (shown inside the dashed lines) is equivalent to an inductor L_a and an ideal transformer T_a with turns ratio of 1. C_{b1} and C_{b2} are divided capacitors with the same voltage: $V_{in}/2$. The voltage across inductor L_a, v_{La}, and the main transformer primary voltage, v_p, are:

$$v_{La} = \left(v_A - \frac{V_{in}}{2}\right) - \left(v_B - \frac{V_{in}}{2}\right) = v_A - v_B = v_{AB} \tag{4.62}$$

$$v_p = \frac{1}{2}\left(v_A - \frac{V_{in}}{2} + v_B - \frac{V_{in}}{2}\right) = \frac{1}{2}(v_A + v_B - V_{in}) \tag{4.63}$$

Obviously, v_{La} is complementary to v_p in term of duty cycle.

In order for the transformer primary winding to obtain a voltage of $(V_{in} - v_A - v_B)/m$, the transformer must be split into two, and the secondary windings of the two

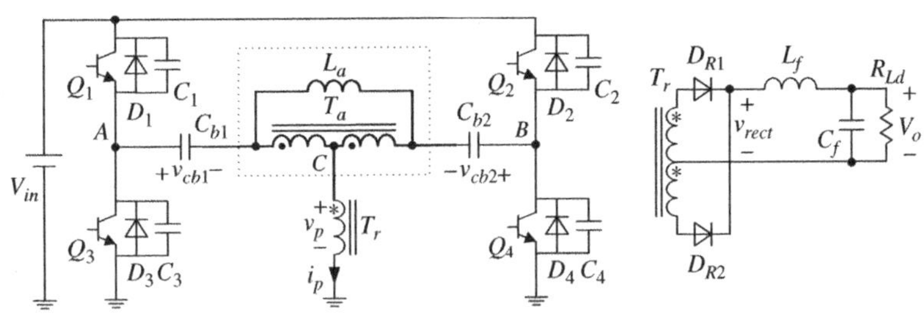

Figure 4.20 Full-bridge converter with transformer primary voltage $(v_A + v_B - V_{in})/2$

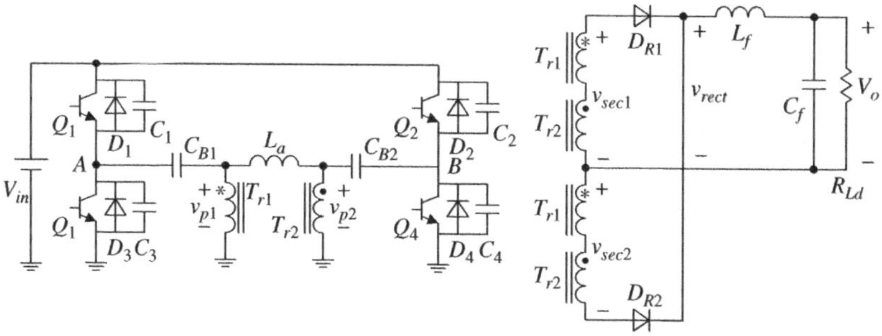

Figure 4.21 Full-bridge converter with two transformers, which achieve an auxiliary inductor current that is adaptive to load

transformers connected in series, as shown in Figure 4.21 [12]. The primary voltages of the two transformers are:

$$v_{p1} = v_A - \frac{V_{in}}{2} \quad (4.64)$$

$$v_{p2} = v_B - \frac{V_{in}}{2} \quad (4.65)$$

And their sum is:

$$v_{p1} + v_{p2} = v_A + v_B - V_{in} \quad (4.66)$$

4.8.5 Auxiliary-Current-Source Networks with Adaptive Resonant Inductor Current

As pointed out in Chapter 3, the lagging leg utilizes the energy stored in the resonant inductor (including the transformer leakage inductor) to achieve ZVS. Reference [13] proposes a ZVS PWM full-bridge converter which resonant inductor current is adaptive to load current, as shown in Figure 4.22. The primary voltage of the auxiliary transformer v_{pa} is approximately equal to $(V_{in} - v_A - v_B)/2$. This voltage is reflected to the secondary side, forcing the resonant inductor current i_{Lr} to increase linearly

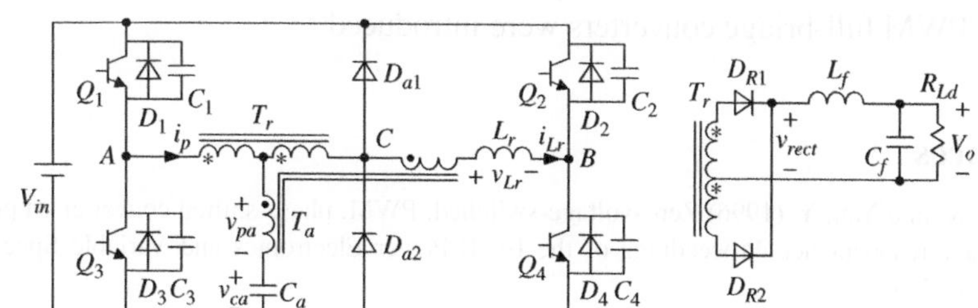

Figure 4.22 Full-bridge converter whose resonant inductor current is adaptive to load

when $v_{AB}=0$. The increase in i_{Lr} is proportional to the duty cycle of the voltage $(V_{in}-v_A-v_B)/2$. At light load, the primary duty cycle is small and the duty cycle of $(V_{in}-v_A-v_B)/2$ is large, so the increase in i_{Lr} is large, which is of benefit in achieving ZVS for the lagging leg. On the other hand, at heavy load, the primary duty cycle is large and the duty cycle of $(V_{in}-v_A-v_B)/2$ is small, so the increase in i_{Lr} is small, but the original i_{Lr} is large enough to achieve ZVS for the lagging leg. Therefore, the auxiliary inductor current is adaptive to the load current, leading to a wide load range for ZVS and reduced conduction loss at heavy load.

4.9 Summary

This chapter described the current-enhancement principle and introduced an auxiliary current source into the full-bridge converter in order to help the resonant inductor achieve ZVS for the lagging leg. As a consequence, the resonant inductor could be reduced, and thus so too could the duty cycle loss. An auxiliary-current-source network consisting of an inductor, two capacitors, and two diodes, based on the current-enhancement principle, was presented, which had the following advantages:

1. The lagging leg could realize ZVS over nearly the entire load range and input voltage range.
2. The secondary duty cycle loss was reduced to nearly zero, which was beneficial to improving efficiency.
3. The auxiliary-current-source network was very simple, and it had no active power switches.
4. The voltage and current stress of the inductor, capacitors, and diodes in the auxiliary-current-source network were very small, and were independent of the load.
5. The auxiliary capacitors did not participate in the switching transition of the lagging leg, making parameter design very easy.

The operating principle and parameter design of the ZVS PWM full-bridge converter with auxiliary-current-source network were presented, and experimental results were provided to verify the theoretical analysis. Various auxiliary-current-source networks for ZVS PWM full-bridge converters were introduced.

References

1. Ruan, X. and Yan, Y. (1996) Zero-voltage-switched, PWM, phase-shifted converter employing an auxiliary resonant net. Proceedings of the IEEE Power Electronics and Variable Speed Drives, pp. 534–539.
2. Hua, G., Lee, F.C. and Jovanovic, M.M. (1991) An improved zero-voltage-switched PWM converter using a saturable inductor. Proceedings of the IEEE Power Electronics Specialists Conference, pp. 189–194.

3. Patterson, O.D. and Divan, D.M. (1987) Pseudo-resonant full bridge dc/dc converter. Proceeding of the IEEE Power Electronics Specialists Conference, pp. 424–430.
4. Xu, M. (1997) Research on PWM soft-switching topology theory. PhD Dissertation, Zhejiang University, Hangzhou, China.
5. Cho, J.G., Sabate, J.A. and Lee, F.C. (1994) Novel full-bridge zero-voltage-transition PWM dc/dc converter for high power applications. Proceedings of the IEEE Applied Power Electronics Conference and Exposition, pp. 143–149.
6. Hamada, S. and Nakaoka, M. (1994) Analysis and design of a saturable reactor assisted soft-switching full-bridge dc-dc converter. *IEEE Trans. Power Electron.*, **9** (3), 309–317.
7. Yang, B., Duarte, J.L., Li, W. *et al.* (2010) Phase-shifted full bridge converter featuring ZVS over the full load range. Proceedings of the IEEE International Conference on Industrial Electronics, pp. 644–649.
8. Borage, M., Tiwari, S. and Kotaiah, S. (2005) A passive auxiliary circuit achieves zero-voltage-switching in full-bridge converter over entire conversion range. *IEEE Trans. Power Electron. Lett.*, **3** (4), 141–143.
9. Chen, Z., Ji, B., Ji, F. and Shi, L. (2010) Analysis and design considerations of an improved ZVS full-bridge dc-dc converter. Proceedings of the IEEE Applied Power Electronics Conference and Exposition, pp. 1471–1476.
10. Jang, Y. and Jovanovic, M.M. (2004) A new family of full-bridge ZVS converters. *IEEE Trans. Power Electron.*, **19** (3), 701–708.
11. Jang, Y., Jovanovic, M.M. and Chang, Y.M. (2003) A new ZVS-PWM full-bridge converter. *IEEE Trans. Power Electron.*, **18** (5), 1122–1129.
12. Borage, M., Tiwari, S., Bhardwaj, S. and Kotaiah, S. (2008) A full-bridge dc–dc converter with zero-voltage-switching over the entire conversion range. *IEEE Trans. Power Electron.*, **23** (4), 1743–1750.
13. Jang, Y. and Jovanovic, M.M. (2007) A new PWM ZVS full-bridge converter. *IEEE Trans. Power Electron.*, **22** (3), 987–994.

5

Zero-Voltage-and-Zero-Current-Switching PWM Full-Bridge Converters

Chapter 2 classified soft-switching pulse-width-modulation (PWM) full-bridge converters into two types, zero-voltage switching (ZVS) and zero-voltage-and-zero-current switching (ZVZCS). For the ZVS type, the zero state operates in current constant mode, and both the leading and the lagging leg realize ZVS. For the ZVZCS type, the zero state operates in current-reset mode, and the leading leg realizes ZVS while the lagging leg realizes zero-current switching (ZCS). This chapter presents the topologies and PWM strategies for ZVZCS full-bridge converters. The methods for resetting the primary current and keeping it at zero in the zero state are described. Consequently, several ZVZCS PWM full-bridge converters are derived, which realize ZVS for the leading leg and ZCS for the lagging leg in a wide load range. A ZVZCS PWM full-bridge converter that employs two diodes in series with the lagging leg is used as an example to illustrate the operating principle. The steady-state analysis and design procedures of the ZVZCS PWM full-bridge converter are presented and the experimental measurements from a 54 V/100 A output prototype are provided.

5.1 Modulation Strategies and Topologies of a ZVZCS PWM Full-Bridge Converter

5.1.1 Modulation of the Leading Leg

Figure 5.1 shows the full-bridge converter operating in zero state, with the power switch Q_4 and antiparallel diode D_3 conducting. As in the ZVS PWM full-bridge converter, capacitors C_1 and C_3 are connected in parallel with the leading-leg switches Q_1 and Q_3, respectively, to realize ZVS. This section looks at the turn-on of Q_1 and Q_3.

Soft-Switching PWM Full-Bridge Converters: Topologies, Control, and Design, First Edition. Xinbo Ruan.

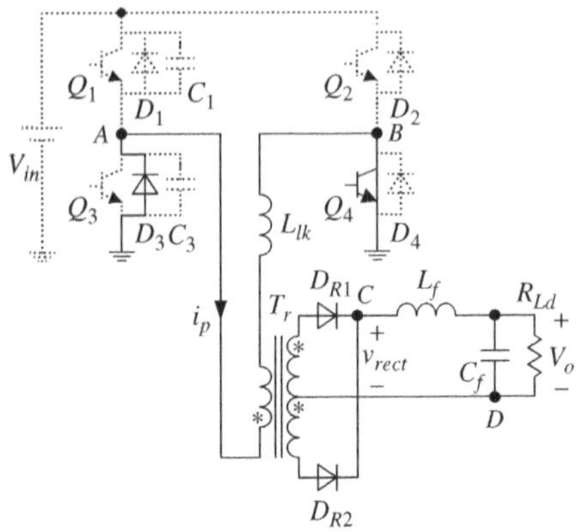

Figure 5.1 Zero state

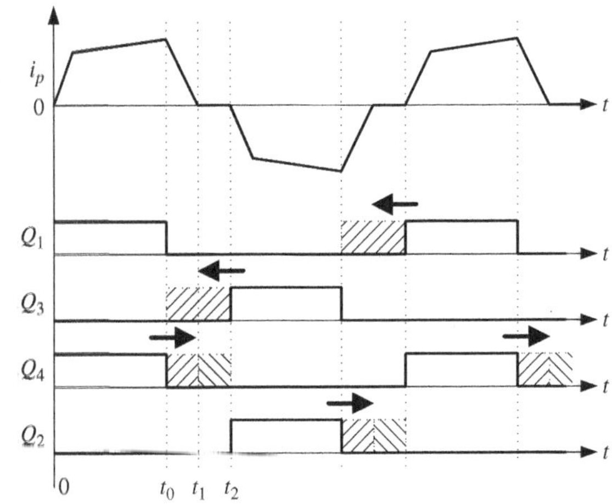

Figure 5.2 Modulation of the leading and the lagging leg to achieve ZVS for the former and ZCS for the latter

Figure 5.2 gives the waveforms of the primary current and modulation of the leading and the lagging leg. The leading-leg power switch Q_1 is turned off at t_0. After fully charging C_1 and discharging C_3, diode D_3 conducts, providing zero-voltage turn-on conditions for the other leading-leg power switch Q_3. From the turn-off instant of Q_1 to the turn-on instant of Q_2 (i.e., during the time interval $[t_0, t_2]$) the full-bridge converter operates in zero state. As pointed out in Chapter 2, in order to achieve ZCS for the lagging leg, the zero state operates in current-reset mode and the primary current i_p is forced to reduce to zero and then stays there (resetting of the primary current will be discussed in Section 5.1.4). When Q_1 is turned off and D_3 conducts, if Q_1 is not turned on in a timely manner then i_p will be forced to zero and D_3 will turn off. C_3 will then be recharged and Q_3 will thus lose the zero-voltage turn-on condition. The time required for i_p to decay to zero is dependent on the load current. So, in order to achieve zero-voltage turn-on for Q_3 under any load, the turn-on instant of Q_3 should be pushed forward to the turn-off instant of Q_1. This means that the on-time of Q_3 must be increased to $T_s/2$, where T_s is the switching period. Likewise, the on-time of Q_1 must also be increased to $T_s/2$, as shown in Figure 5.2. In practice, there should be a dead time between the drive signals of Q_1 and of Q_3 in order to let the parallel capacitor of the incoming switch be completely discharged and make the antiparallel diode conduct to provide zero-voltage turn-on conditions for the incoming switch.

5.1.2 Modulation of the Lagging Leg

Since the lagging-leg switches are intended to realize ZCS, they cannot have parallel-connected capacitors. With reference to Figure 5.2, after the leading-leg power switch Q_1 is turned off, the full-bridge converter enters into the zero state and the primary current i_p begins to decay, reaching zero at t_1. Thus, the lagging-leg switch Q_4 can be turned off at zero current after t_1. From t_1 to t_2, i_p is kept at zero. The turn-off instant of Q_4 can thus also be pushed backward to t_2, maintaining zero-current turn-off. The other lagging-leg power switch Q_2 achieves zero-current turn-off in a similar way.

In conclusion, the leading leg has two kinds of on-time arrangement: one is to adjust the turn-off instant of the lagging leg backward to allow the primary current to decay to zero; the other is to adjust the turn-off instant of the lagging leg backward to let the on-time be equal to $T_s/2$.

5.1.3 Modulation Strategies of ZVZCS PWM Full-Bridge Converters

As discussed, there is only one kind of on-time for the leading leg; that is, the turn-on instant of the leading leg should be pushed forward to let the on-time equal $T_s/2$. There are two kinds of on-time for the lagging leg: pushing the turn-off instant of the lagging leg backward to allow the primary current to decay to zero and pushing the turn-off instant of the lagging leg backward to let the on-time equal $T_s/2$. Hence, there are two kinds of modulation strategy for the ZVZCS PWM full-bridge converters: strategies 6 and 9 in Chapter 2, which are shown in Figure 5.3.

5.1.4 Method for Resetting the Primary Current at Zero State

5.1.4.1 Blocking the Voltage Source

In the zero state, both rectifier diodes are conducting and both the primary and the secondary voltages of the transformer are zero. In order to make the primary current i_p decay to zero in the zero state, a blocking voltage source v_{block} can be inserted in series

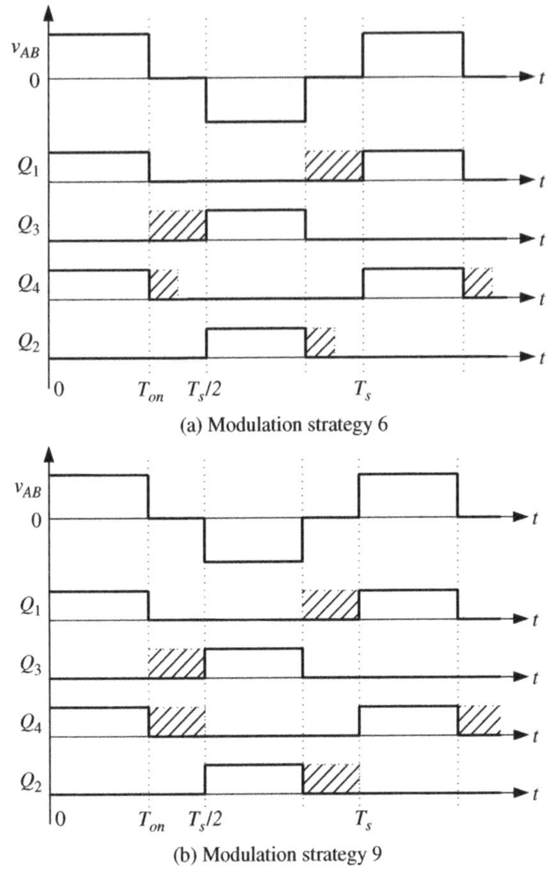

Figure 5.3 PWM strategies suitable for ZVZCS full-bridge converters

with the primary winding, as shown in Figure 5.4a. For brevity of illustration, the circuit of the secondary part is omitted in Figure 5.4. In the zero state, if i_p flows in the positive direction then the blocking voltage source is positive, as shown in Figure 5.4b; if i_p flows in the negative direction, the blocking voltage source is negative, as shown in Figure 5.4c. The blocking voltage source can be simply realized by a capacitor C_b, called blocking capacitor, as shown in Figure 5.5a. When Q_1 and Q_4 conduct, i_p charges C_b; when Q_2 and Q_3 conduct, i_p discharges C_b. In the zero state, the voltage of C_b remains constant and resets i_p, as shown in Figure 5.5b.

5.1.4.2 Blocking the Primary Current Reverse Flow Path

In the zero state, in order to achieve ZCS for the lagging leg the primary current i_p should be prevented from turning negative after it decays to zero. Therefore, the reverse path for i_p should be blocked. As can be seen from Figure 5.5a, there are three

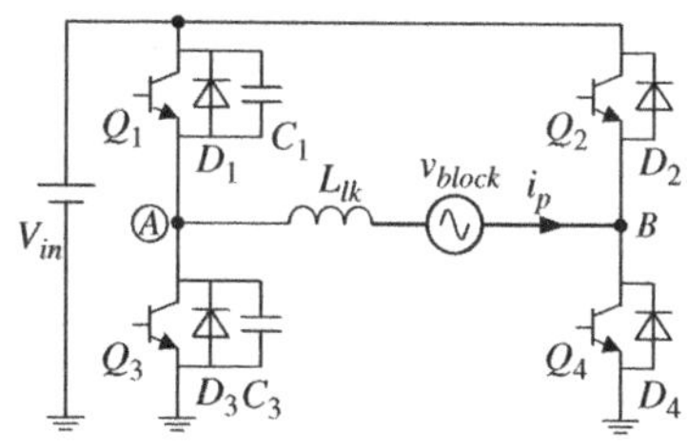

(a) Adding the blocking voltage source

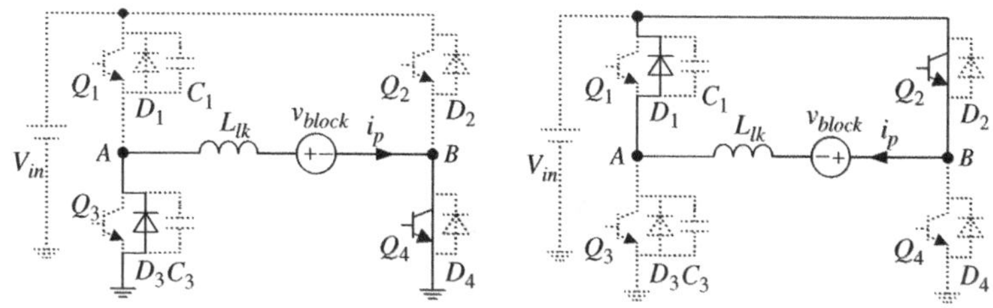

(b) $i_p > 0$ (c) $i_p < 0$

Figure 5.4 Requirement of the blocking voltage source

possible places at which to block the reverse path: (i) segment AO/AC; (ii) segment AB; and (iii) segment BO/BC.

As illustrated in Section 5.1.1, in order to achieve ZVS for the leading leg, either Q_1 or Q_3 is turned on during the zero state, which makes it impossible to prevent i_p from flowing in the reverse direction in segment AO/AC.

In segment AB, there are several methods for blocking the reverse flow path:

1. Adding a saturable inductor, as shown in Figure 5.6a [1]. In the zero state, the saturable inductor works under linear conditions and behaves as a high-impedance device to prevent i_p from flowing in the reverse direction. As in the +1 or −1 state, the saturable inductor is saturated.
2. Adding an active clamping circuit consisting of Q_b, D_b, and C_b into the secondary side, as shown in the dashed block in Figure 5.6b [2].
3. Adding an auxiliary current-reset circuit into the secondary side. Figure 5.6c,d shows two auxiliary current-reset circuits, one consisting of D_{b1}, D_{b2}, and C_b [3] and the other of D_{b1}, D_{b2}, D_{b3}, C_{b1}, and C_{b2} [4].
4. Adding two auxiliary transformer windings, followed by a full-wave rectifier and a filter capacitor, as shown in the dashed block in Figure 5.6e [5].

In all the methods for resetting the primary current shown in Figure 5.6b–e, the blocking capacitor is moved from the primary to the secondary side and an associated circuit is added. This decouples the blocking capacitor from the primary side when the primary current decays to zero, meaning the blocking capacitor voltage cannot force the primary current to increase in the reverse direction.

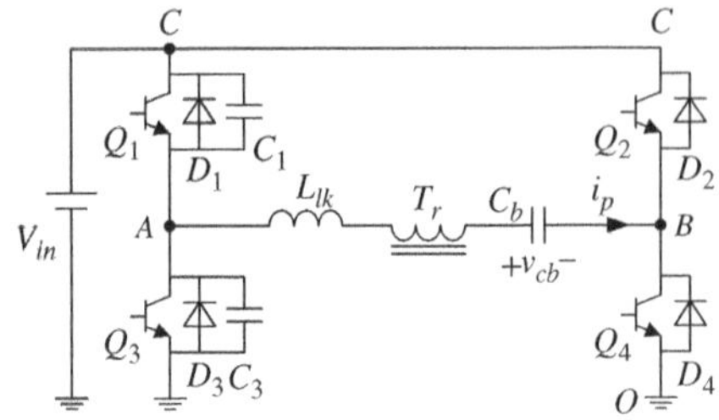

(a) Main circuit

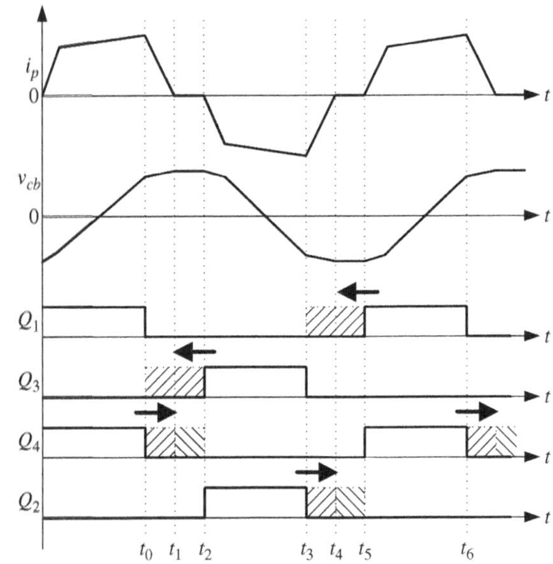

(b) Key waveforms

Figure 5.5 Realization of the blocking voltage source and the key waveforms

In segment BO/BC, we can add two diodes D_2 and D_4 in series with the lagging-leg switches Q_2 and Q_4, respectively, to make Q_2 and Q_4 conduct only in the positive direction. A ZVZCS PWM full-bridge converter is thus constructed, as shown in Figure 5.6f [6].

5.1.4.3 Comparison of ZVZCS PWM Full-Bridge Converters

The converter in Figure 5.6a is very simple, as only one saturable inductor is needed. However, the use of a saturable inductor has two drawbacks. First, it causes a secondary duty cycle loss. In general, the input voltage V_{in} has a certain variation range, and L_s should be designed according to $V_{in\max}$ to prevent its saturation during the zero state. Thus, the effective secondary duty cycle is reduced as V_{in} decreases. The lower the input voltage V_{in}, the larger the secondary duty cycle loss. Second, it causes high core loss, since it saturates in both directions.

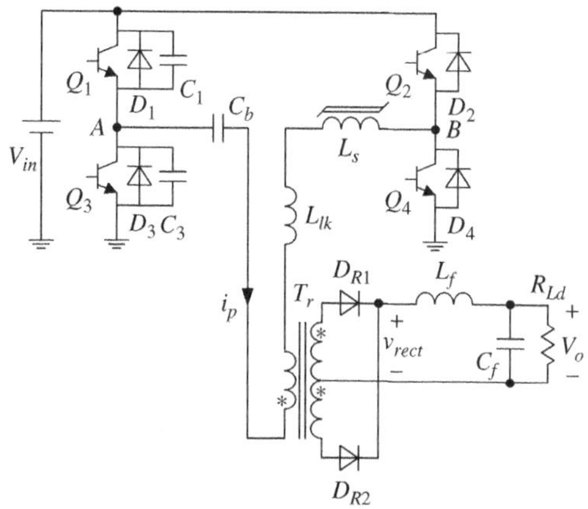

(a) Using the saturable inductor

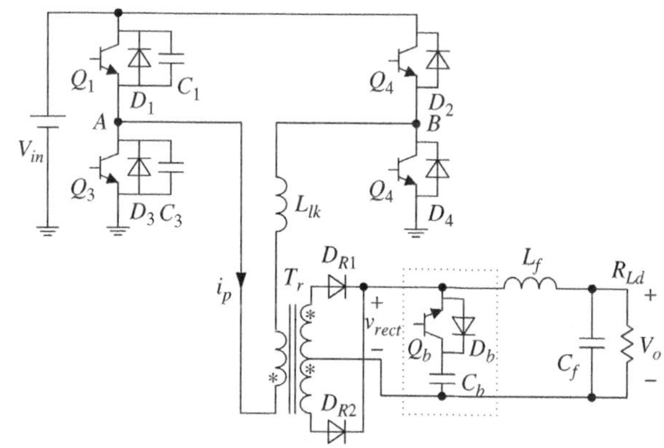

(b) Using the active clamp circuit

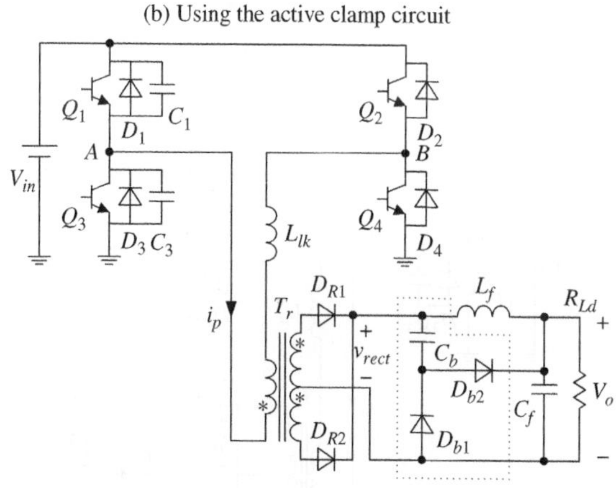

(c) No.1 secondary reset circuit

Figure 5.6 Topologies of ZVZCS PWM full-bridge converters

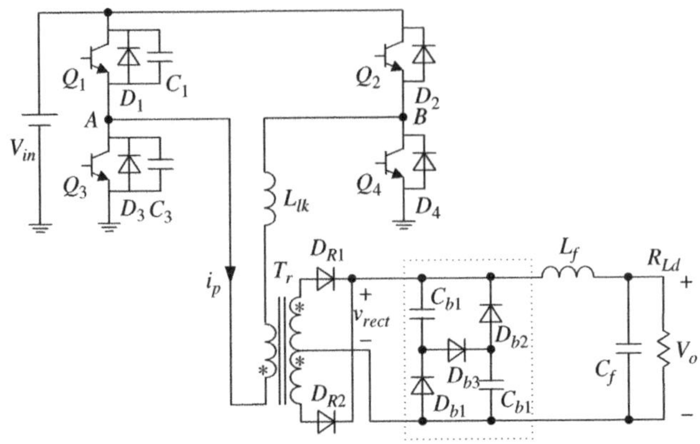

(d) No.2 secondary reset circuit

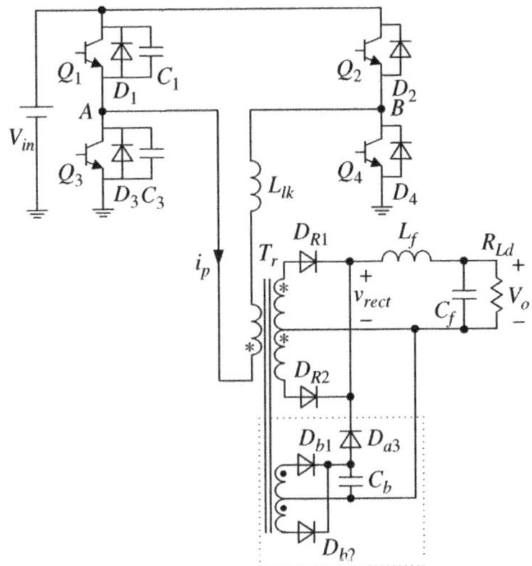

(e) Adding an auxiliary winding

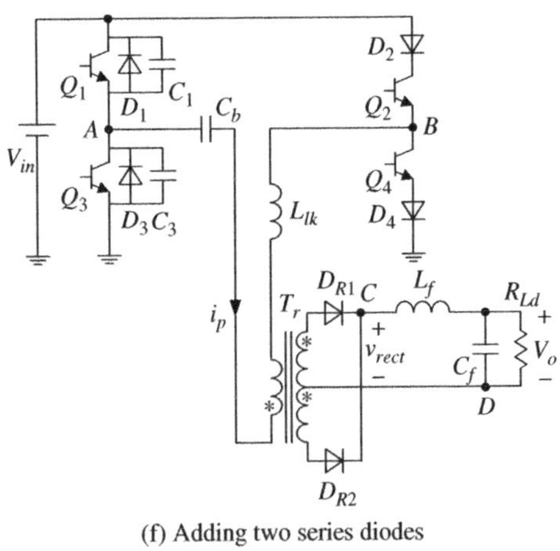

(f) Adding two series diodes

Figure 5.6 (*Continued*)

In Figure 5.6b, the blocking capacitor is shifted to the secondary side and an active switch is introduced, which not only resets the primary current but also clamps the rectified voltage to suppress the parasitic voltage oscillation. However, the converter requires an additional active switch and the associated control circuitry.

In Figure 5.6c–e, the blocking capacitor is also shifted to the secondary side. The primary current-reset mechanisms of all three converters are similar, and the energy stored in the blocking capacitor is delivered to the load during zero state. The converter in Figure 5.6c is better than the other two in term of simplicity. Furthermore, the peak secondary rectified voltage is inversely proportional to the duty cycle and is less than twice the reflected input voltage. For example, if the duty cycle is 0.8, the peak rectified voltage is 1.2 times the input voltage reflected to the secondary side. However, this converter also has three drawbacks:

1. In the zero state, the blocking capacitor is completely discharged. When the lagging-leg switch turns on, the primary current has an excess high current spike due to charging of the blocking capacitor. The larger the blocking capacitor, the larger the current spike.
2. When the leading-leg switch turns off, the rectified voltage ramps down. As it becomes lower than the blocking capacitor voltage, the blocking capacitor supplies the rectified output filter inductor current. Thus, only the energy stored in the leakage inductor (instead of the output filter inductor) is used to charge and discharge the parallel-connected capacitors of the leading-leg switches. In general, the leakage inductor is too small to completely discharge the remaining charge of the parallel capacitor of the incoming leading-leg switch, which may result in hard turn-on of the leading-leg switch.
3. As the converter operates in the zero state, the blocking capacitor not only needs to reset the primary current but also provides the rectified output current. If the blocking capacitor is not sufficiently large, its energy will be insufficient to reset the primary current.

The proposed novel converter in Figure 5.6f introduces just two diodes and does not have any of these drawbacks. When the leading-leg switch turns on, there is no excess current spike because the blocking capacitor is in series with the load (including L_f, C_f, and R_{Ld}), instead of being in parallel with it. During the switching interval of the leading leg, the output filter inductor is in series with the leakage inductor, and the energy stored in both is used to achieve ZVS for the leading leg. Thus, it is easy to realize ZVS. In the zero state, the output current freewheels via the two secondary rectifier diodes; the voltage of the blocking capacitor is fully used to reset the primary current. As will be explained later, the lagging leg can achieve ZCS under any load condition. One drawback of this converter is that the primary-side conduction losses are slightly increased due to the presence of the series diodes. However, these increased losses are not very significant.

5.2 Operating Principle of a ZVZCS PWM Full-Bridge Converter

Section 5.1 discussed PWM strategies and topologies for ZVZCS full-bridge converters. In this section, the operating principle of the novel ZVZCS PWM full-bridge converter (Figure 5.6f) employing a phase-shifted modulation strategy is presented. Figure 5.7 shows the key waveforms. The following assumptions are made in this analysis: (i) all power switches and diodes are ideal; (ii) all capacitors and inductors are ideal; (iii) the blocking capacitor C_b is sufficiently large; (iv) $C_1 = C_3 = C_r$; and (v) $L_{lk} \ll K^2 L_f$, where K is the primary-to-secondary-windings-turns ratio of the transformer.

During a switching period, there are 10 switching modes (equivalent circuits are shown in Figure 5.8). Since the first half-period is similar to the second, only the first is given here for brevity, which includes 5 switching modes and 1 initial switching mode:

1. **Mode 0, prior to t_0 (Figure 5.8a):** Prior to t_0, Q_1 and Q_4 are conducting and the primary current i_p charges the blocking capacitor C_b. $I_p(t_0) = I_{p0} = I_o/K$, where I_o is the output current and $V_{cb}(t_0)$ is the voltage of C_b at t_0.
2. **Mode 1, $[t_0, t_1]$ (Figure 5.8b):** Q_1 is turned off at t_0 and i_p charges C_1 and discharges C_3. Q_1 is turned off with ZVS because C_3 and C_1 limit the rising rate of the voltage across it. During this interval, the transformer leakage inductor L_{lk} is

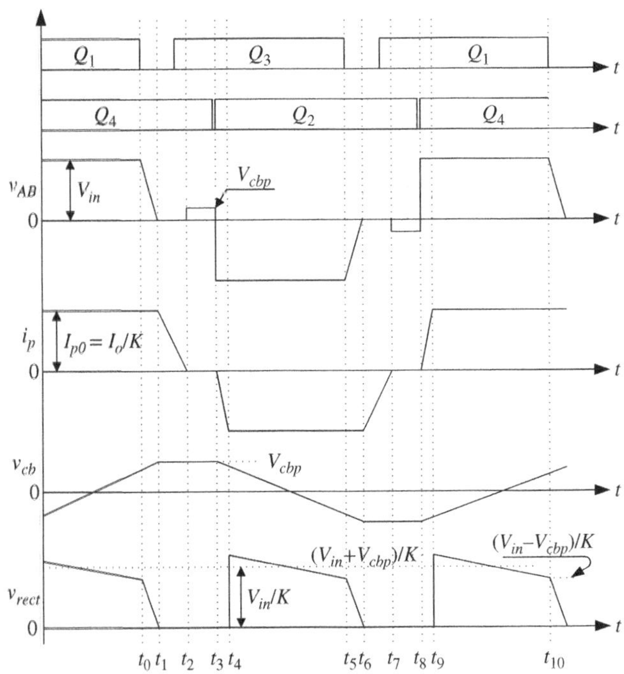

Figure 5.7 Key waveforms of the novel ZVZCS PWM FB converter

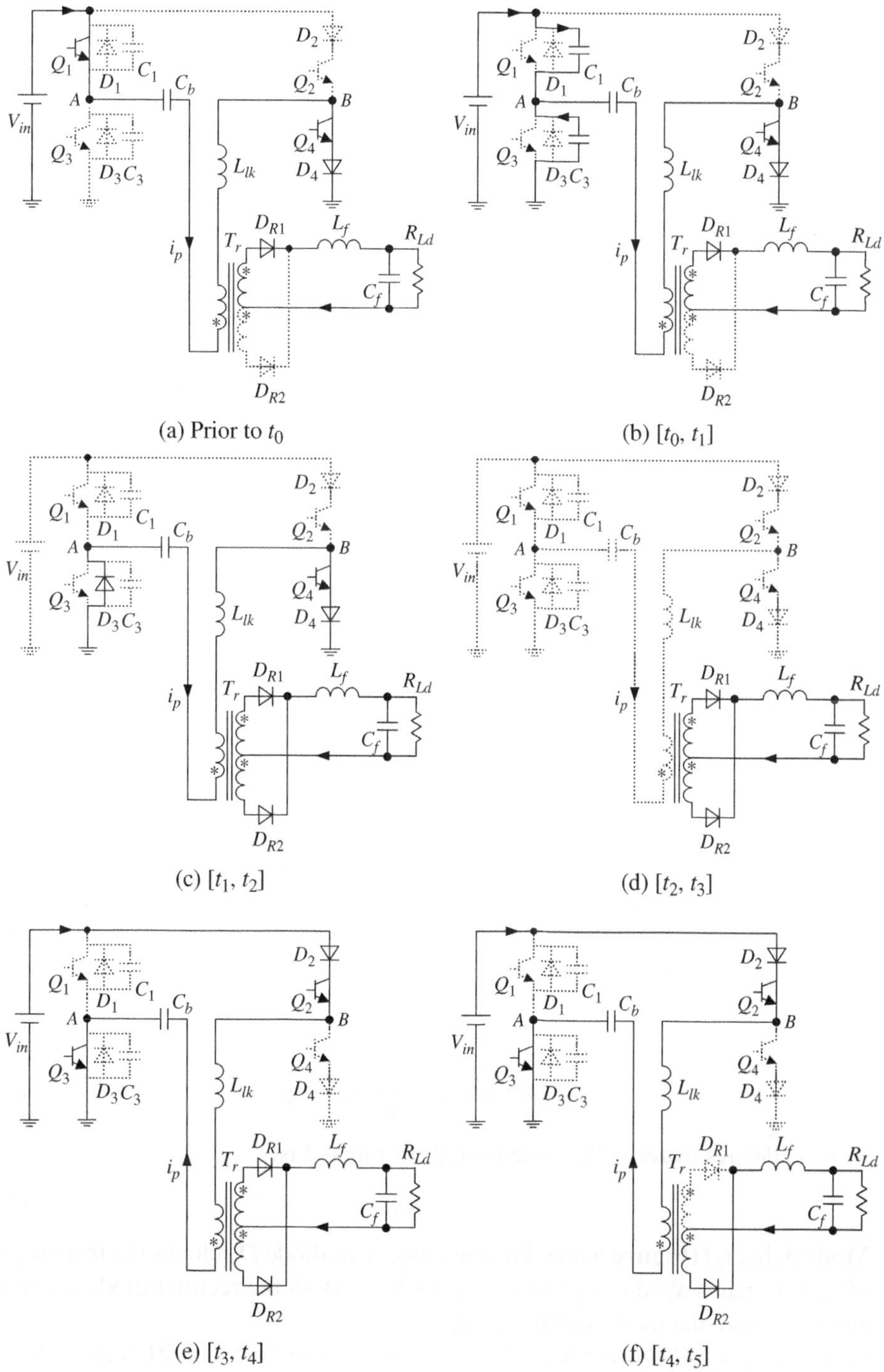

Figure 5.8 Equivalent circuits of the different switching modes

in series with L_f, which is large enough to be treated as a current constant source, so that i_p keeps the value $I_{p0} = I_o/K$. i_p continues charging C_b. The voltage of C_1 rises linearly and the voltage of C_3 decays linearly. The voltages of C_b, C_1, and C_3 are given by:

$$v_{Cb}(t) = V_{Cb}(t_0) + I_{p0}\frac{t - t_0}{C_b} \tag{5.1}$$

$$v_{C1}(t) = \frac{I_{p0}}{2C_r}(t - t_0) \tag{5.2}$$

$$v_{C3}(t) = V_{in} - \frac{I_{p0}}{2C_r}(t - t_0) \tag{5.3}$$

At t_1, v_{C3} decays to zero, D_3 turns on naturally, and mode 1 ends. The time period for mode 1 is:

$$t_{01} = 2C_r V_{in}/I_{p0} \tag{5.4}$$

At t_1, the voltage of C_b is:

$$V_{Cb}(t_1) = V_{Cb}(t_0) + 2\frac{C_r V_{in}}{C_b} \tag{5.5}$$

3. **Mode 2, [t_1, t_2] (Figure 5.8c):** As D_3 is conducting, Q_3 is turned on with ZVS. The dead time between the drive signals of Q_1 and Q_3 should be greater than the time period of mode 1; that is:

$$t_{d(lead)} > 2C_r V_{in}/I_{p0} \tag{5.6}$$

During this interval, D_3 and Q_4 conduct and $v_{AB} = 0$. v_{Cb} is applied to L_{lk} and primary winding, which forces i_p to decrease. i_p is thus insufficient to provide the load current, and the secondary rectifier diodes D_{R1} and D_{R2} conduct simultaneously, clamping both the primary and the secondary voltage of the transformer at zero. Thus, v_{Cb} is fully applied on L_{lk}. Since C_b is large enough to be treated as a constant voltage source during this mode, i_p decays linearly; that is:

$$v_{Cb}(t) = V_{Cb}(t_1) \equiv V_{Cbp} \tag{5.7}$$

$$i_p(t) = I_{p0} - \frac{V_{Cbp}}{L_{lk}}(t - t_0) \tag{5.8}$$

At t_2, i_p decays to zero. The time period for mode 2 is:

$$t_{12} = L_{lk} I_{p0}/V_{Cbp} \tag{5.9}$$

4. **Mode 3, [t_2, t_3] (Figure 5.8d):** During mode 3, as diode D_4 blocks the reverse path of i_p, i_p remains at zero. $v_A = 0$ and $v_B = -V_{Cbp}$. Both the rectifier diodes conduct, and they share the load current equally.
5. **Mode 4, [t_3, t_4] (Figure 5.8e):** As $i_p = 0$, Q_4 is turned off with ZCS at t_3. After a very short delay, Q_2 is turned on with ZCS, because L_{lk} limits di_p/dt. As the primary

current i_p reflected to the secondary side is lower than the load current, both the secondary rectifier diodes conduct, clamping both the primary and the secondary voltages of the transformer at zero. Thus, $-(V_{in}+V_{Cbp})$ is applied to L_{lk}, causing i_p to rise linearly in the negative direction; that is:

$$i_p(t) = -\frac{V_{in}+V_{Cbp}}{L_{lk}}(t-t_3) \tag{5.10}$$

At t_4, i_p rises to the reflected load current. The time interval of this mode is:

$$t_{34} = \frac{L_{lk}I_{p0}}{V_{in}+V_{Cbp}} \tag{5.11}$$

6. **Mode 5, [t_4, t_5] (Figure 5.8f):** From t_4, the primary side powers the load and i_p discharges C_b. D_{R1} turns off naturally and D_{R2} carries all the output current. The voltage of C_b is given by:

$$v_{Cb}(t) = V_{Cbp} - \frac{I_{p0}}{C_b}(t-t_4) \tag{5.12}$$

At t_5, Q_3 is turned off and the second half-cycle [t_4, t_{10}] begins; the process is similar to the first half-cycle [t_0, t_5]. At this time, the voltage of C_b is:

$$V_{Cb}(t_5) = V_{Cbp} - \frac{I_{po}}{C_b}t_{45} \tag{5.13}$$

5.3 Theoretical Analysis

5.3.1 *Peak Voltage of the Block Capacitor*

From Equation 5.9, it is clear that the interval t_{12} is determined by V_{Cbp}, so it is important to calculate this value. v_{Cb} reaches its negative peak value $-V_{Cbp}$ at t_6. The operation during [t_5, t_6] is similar to that during [t_0, t_1], giving the following equation:

$$V_{Cb}(t_6) = V_{Cb}(t_5) - 2\frac{C_rV_{in}}{C_b} = V_{Cbp} - \frac{I_{p0}}{C_b}t_{45} - 2\frac{C_rV_{in}}{C_b} = -V_{Cbp} \tag{5.14}$$

In general, $C_r \ll C_b$. Equation 5.14 can thus be simplified to:

$$V_{cbp} = \frac{I_{p0}}{2C_b}t_{45} \tag{5.15}$$

5.3.2 *Achieving ZVS for the Leading Leg*

Clearly, ZVS of the leading leg is realized by utilizing the energy stored in the output filter inductor. In general, the output filter inductor is large enough that ZVS of the leading leg can be achieved over a wide load range.

5.3.3 Maximum Effective Duty Cycle

In order to achieve ZCS for the lagging leg, i_p should decay to zero before the switching of the lagging leg. t_{12} can be derived from Equations 5.9 and 5.15 as:

$$t_{12} = \frac{2L_{lk}C_b}{t_{45}} = \frac{2L_{lk}C_b}{D_{eff}T_s/2} = \frac{4L_{lk}C_b}{D_{eff}T_s} \tag{5.16}$$

where D_{eff} is the effective duty cycle and T_s is the switching period.

According to Figure 5.9, D_{effmax} is determined by:

$$D_{effmax} < 1 - D_{reset} - D_{ZCS} - D_{loss} \tag{5.17}$$

where $D_{reset} = \frac{t_{12}}{T_s/2}$ and $D_{ZCS} = \frac{t_{ZCS}}{T_s/2}$. ZCS is achieved during interval T_{ZCS}, which is determined by the turn-off characteristics of the switching device, such as the minority carrier life time of insulated gate bipolar transistors (IGBTs). $D_{loss} = \frac{t_{34}}{T_s/2}$ is the duty cycle loss caused by the leakage inductor.

5.3.4 Achieving ZCS for the Lagging Leg

From Equation 5.16, it can be seen that t_{12} is independent of load current and is inversely proportional to D_{eff}. If Equation 5.17 is satisfied, the lagging leg can achieve ZCS over the entire line and load range.

5.3.5 Voltage Stress of the Lagging Leg

During mode 3, $i_p = 0$, $v_B = -V_{Cbp}$, the lagging-leg voltage is:

$$V_{Q2} = V_{in} + V_{cbp} \tag{5.18}$$

$$V_{Q4} = -V_{cbp} \tag{5.19}$$

From Equations 5.18 and 5.19, it can be seen that the maximum voltage across the lagging-leg switches is $V_{in} + V_{Cbp}$. The negative voltage $-V_{Cbp}$ can be applied to these. Therefore, series diodes are needed to prevent reverse breakdown.

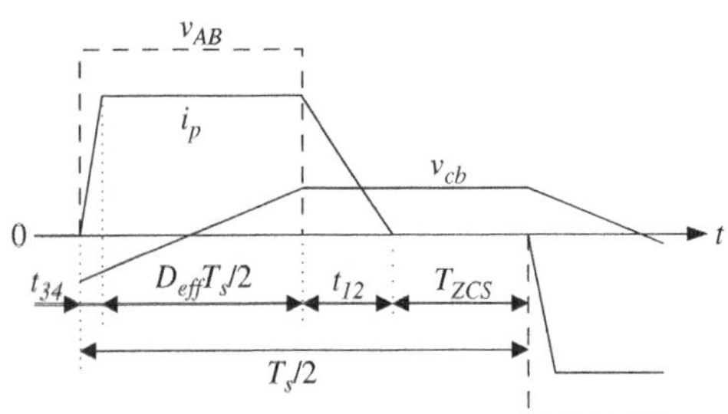

Figure 5.9 Determination of D_{effmax}

5.3.6 *Blocking Capacitor*

The value of the blocking capacitor C_b can be determined by considering two factors: (i) in order to increase D_{effmax}, C_b should be minimized according to Equations 5.16 and 5.17; and (ii) in order to reduce the voltage stress of the lagging leg and the series diodes, C_b should be selected to be as large as possible. There is thus a trade-off in selecting the C_b value. In general, C_b is designed to keep its peak voltage at around 10% of V_{in}; that is, $V_{Cbp} = 10\% V_{in}$.

5.4 Simplified Design Procedure and Example

This section illustrates the design procedure with an example. The design of this converter involves complex interactions between circuit parameters and operating conditions. Some assumptions and approximations are necessary to simplify the design.

The input specifications are as follows:

- input voltage: $V_{in} = 537$ Vdc $\pm 20\%$;
- output voltage: $V_o = 54$ Vdc;
- output current: $I_o = 100$ A;
- switching frequency: $f_s = 25$ kHz; switching period: $T_s = 40$ μs;
- leakage inductor measured at the switching frequency: $L_{lk} = 5$ μH.

5.4.1 *Transformer Winding-Turns Ratio*

The effective duty cycle is maximal at the lowest input voltage and the primary-to-secondary-winding-turns ratio of the transformer is:

$$K = \frac{V_{inmin}}{(V_o + V_D)/D_{effmax}} \tag{5.20}$$

where V_D is the voltage drop in the secondary rectifier diode. Here, $V_D = 1.5$ V and $V_{inmin} = 537 \times (1-20\%) = 429.6$ V. Let $D_{effmax} = 0.7$ at the lowest input voltage; according to Equation 5.20, the transformer turns ratio is then calculated as $K = 5.42$. In practice, the numbers of turns of the secondary and primary windings are selected at 4 and 22, respectively. Thus, $K = 5.5$ and $D_{effmax} = 0.71$.

5.4.2 *Calculation of Blocking Capacitance*

The peak voltage of the blocking capacitor V_{Cbp} is selected to be $10\% V_{in}$; that is, $V_{Cbp} = 10\% V_{in}$. According to Equation 5.15, the blocking capacitance is calculated as:

$$C_b = \frac{I_{p0}}{2V_{cbp}} t_{45} = \frac{I_o/K}{2\times 0.1 \cdot V_{in}} D_{effmax} \frac{T_s}{2} = \frac{100/5.5}{2\times 0.1\times 537} \cdot 0.71 \cdot \frac{40\times 10^{-6}}{2} = 2.4\,(\mu\text{F}) \tag{5.21}$$

We choose $C_b = 2.2$ μF (930C2W2P2K from CDE Co.), so $V_{Cbp} = 58.7$ V.

5.4.3 Verification of the Transformer Turns Ratio and Blocking Capacitance

Once K and C_b are determined, D_{eff}, D_{reset}, and D_{loss} can be calculated from:

$$D_{eff} = \frac{K(V_o + V_D)}{V_{in}} \tag{5.22}$$

$$D_{reset} = \frac{t_{12}}{T_s/2} = \frac{8L_{lk}C_b}{D_{eff}T_s^2} = \frac{8V_{in}L_{lk}C_b}{K(V_o + V_D)T_s^2} \tag{5.23}$$

$$D_{loss} = \frac{t_{34}}{T_s/2} = \frac{2L_{lk}I_o}{KT_s(V_{in} + V_{cbp})} = \frac{2L_{lk}I_o}{KT_s\left(V_{in} + \frac{I_o}{2C_b}\frac{V_o + V_D}{V_{in}}\frac{T_s}{2}\right)} \tag{5.24}$$

In order to meet the voltage and current requirements, IGBT Module (VII50-12Q3) is chosen. The current tail time is $T_{tail} = 0.35\,\mu s$. Therefore, $T_{ZCS} = 0.35\,\mu s$ and $D_{ZCS} = 0.0175$.

We can define D_{sum} as:

$$D_{sum} = D_{eff} + D_{reset} + D_{loss} + D_{ZCS} \tag{5.25}$$

D_{sum}, D_{eff}, D_{reset}, and D_{loss} (Figure 5.10) are all dependent on the input voltage V_{in}. D_{eff} and D_{sum} achieve their maximum values at the minimum input voltage; that is, $D_{effmax} = 0.71$ and $D_{summax} = 0.82 < 1$, satisfying Equation 5.17. We can therefore conclude that the values selected for K and C_b are reasonable.

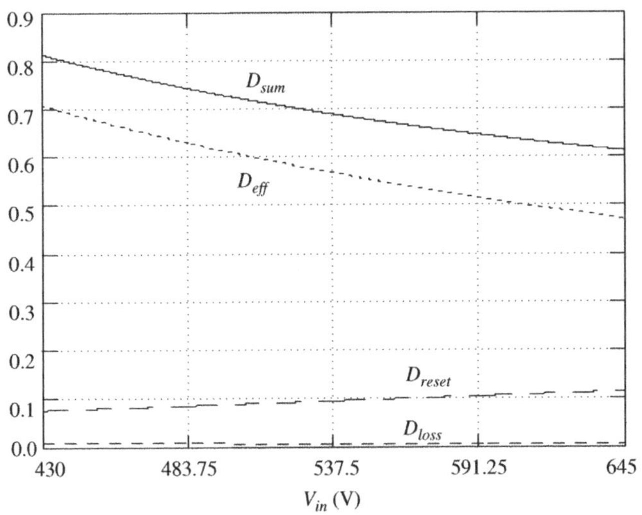

Figure 5.10 D_{sum}, D_{eff}, D_{reset}, and D_{loss} as functions of the input voltage V_{in}

Since Equation 5.17 is satisfied, the lagging leg can achieve ZCS over the entire range of lines and loads.

5.4.4 *Dead Time between the Gate Drive Signals of the Leading Leg*

In order to achieve soft turn-off for the leading leg, snubber capacitors are added in parallel with the switches. Based on the operation analysis in mode 1, t_{01} is the time it takes for the snubber capacitor voltage to rise from zero to V_{in}. In order to reduce the turn-off loss, t_{01} is selected to be (2–3)T_{tail} at full load. According to Equation 5.4, we have:

$$C_1 = C_3 = C_r = \frac{\frac{I_o}{K} \cdot 3 \cdot T_{tail}}{2V_{in}} = 17.8\ (\text{nF}) \tag{5.26}$$

We choose $C_1 = C_3 = C_r = 15$ nF.

In order to ensure zero-voltage turn-on of the leading-leg switches, the voltage of the paralleled capacitor of the incoming leading-leg switch should decay to zero prior to its drive signal stepping to high level. The dead time between the drive signals of the leading-leg switches is selected as $t_d = 2.4\ \mu s$. As can be seen from Equation 5.4, as the load current increases, t_{01} decreases and it becomes easier for the leading-leg switches to achieve ZVS. Thus, according to Equation 5.4, the minimum load current required for the leading leg to realize ZVS is:

$$I_{omin} = K\frac{2C_r V_{in}}{t_d} = 5.5 \times \frac{2 \times 15 \times 10^{-9} \times 537}{2.4 \times 10^{-6}} = 37\ (\text{A}) \tag{5.27}$$

which represents 37% of the full load current.

5.5 Experimental Verification

Following the design guidelines outlined in the previous section, a 54 V/100 A prototype was developed. The following parameters were used in the experiment: input voltage $V_{in} = 537$ Vdc; output voltage $V_o = 54$ Vdc; output current $I_o = 100$ A; transformer turns ratio $K = 5.5$; leakage inductor $L_{lk} = 5\ \mu H$; block capacitor $C_b = 2.2\ \mu F$; snubber capacitor $C_1 = C_3 = C_r = 15$ nF; output inductor $L_f = 30\ \mu H$; output capacitor $C_f = 10\,000\ \mu F$; power switches = IGBT (VII50-12Q3); series diode = DSEP2x31-03A; rectifier diode = MEK95-06DA;switching frequency $f_s = 25$ kHz.

Figure 5.11 shows the experimental waveforms at full load. Figure 5.11a shows the waveforms of v_{AB} and i_p, from which it can be seen that when $v_{AB} = 0$, the blocking capacitor voltage forces i_p to decay to zero, achieving ZCS for the lagging leg. Compared with a ZVS PWM full-bridge converter, a ZVZCS PWM full-bridge converter has no idle current during the zero state, which helps increase its efficiency. Figure 5.11b shows the waveform of v_{AB} and the primary voltage of transformer v_p. v_p is not a square wave, due to the existence of the blocking capacitor. However, its average value is the same as that of a ZVS PWM full-bridge converter. Figure 5.11c

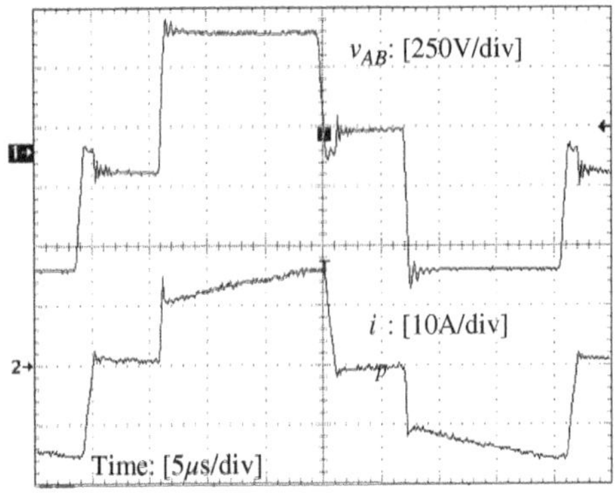

(a) v_{AB} and i_p

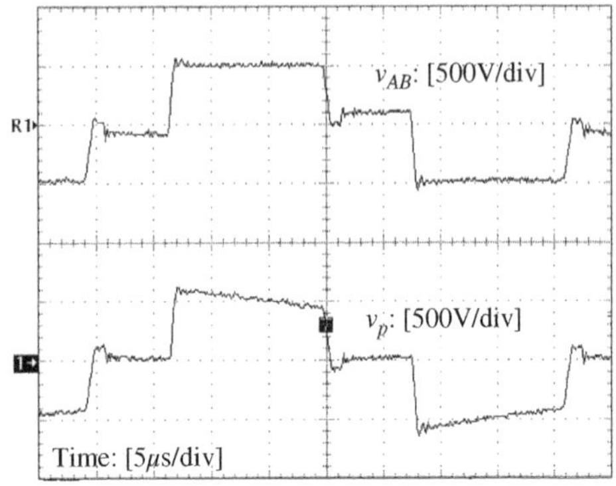

(b) v_{AB} and v_p

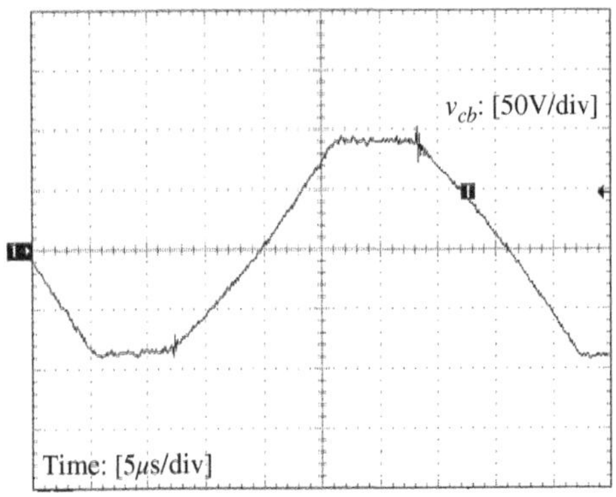

(c) v_{Cb}

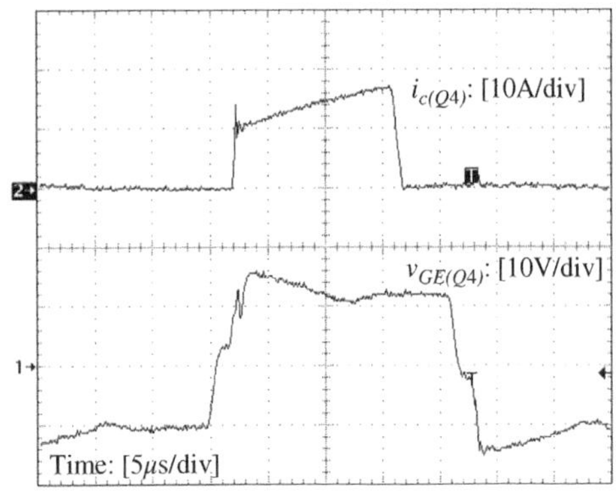

(d) Current and drive signal of Q_4

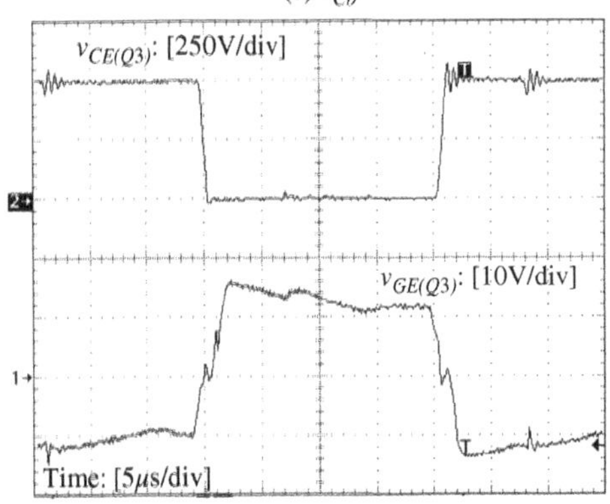

(e) Voltage and drive signal of Q_3

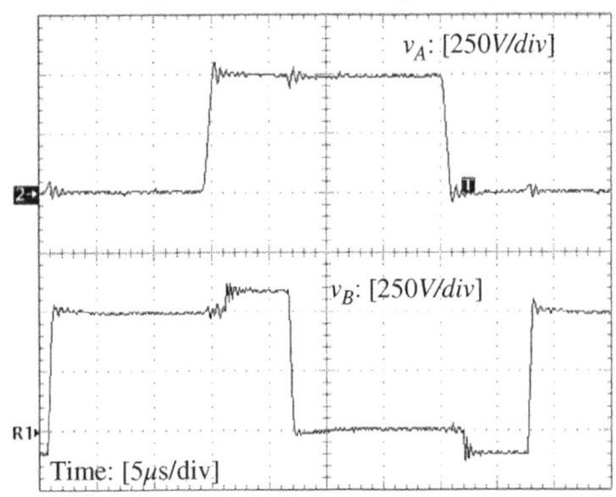

(f) v_A and v_B

Figure 5.11 Experimental results

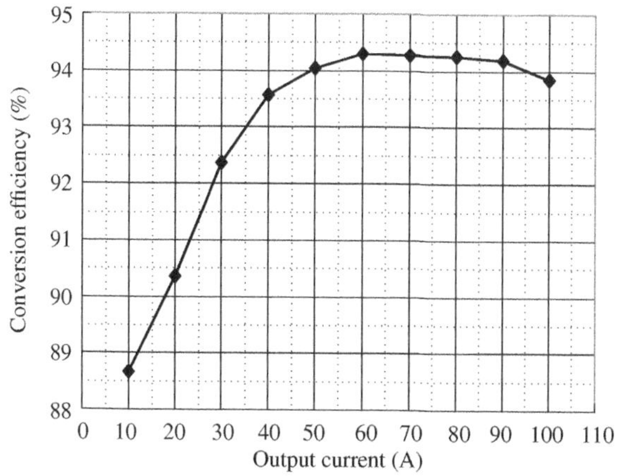

Figure 5.12 Conversion efficiency

shows the waveform of the blocking capacitor voltage and Figure 5.11d shows the waveforms of the collector current and the drive signal of the lagging-leg switch Q_4; it can be seen that Q_4 operates with ZCS. Figure 5.11e shows the waveforms of the collector-to-emitter voltage and the drive signal of the leading-leg switch Q_3; it can be seen that Q_3 operates with ZVS. Figure 5.11f shows the waveforms of the voltages of midpoints of the leading and the lagging leg. As can be seen, the peak voltage across the lagging-leg switch is $V_{in} + V_{Cbp}$ and a negative voltage $-V_{Cbp}$ applies on the lagging-leg switch, so that a series diode is needed to prevent reverse breakdown.

Figure 5.12 gives the overall conversion efficiency at 380 Vac input voltage (which is rectified and filtered to the dc voltage $V_{in} = 537$ Vdc). The efficiency is 93.8% at full load. If a ZVS PWM full-bridge converter is employed, the efficiency is 92% at full load. The higher efficiency is achieved because there is no idle current and no conduction loss in the power switches and transformers during the zero state in the ZVZCS PWM full-bridge converter.

5.6 Summary

This chapter discussed PWM strategies for the ZVZCS full-bridge converter, selecting two suitable ones. Various methods of resetting the primary current and keeping it at zero state were expounded and several ZVZCS PWM full-bridge converters were derived, among which the ZVZCS PWM full-bridge converter employing two additional diodes in series with the lagging leg was shown to have unique advantages: its leading leg can achieve ZVS over a wide load range and its lagging leg can achieve ZCS over the entire range of lines and loads. The operating principle, steady-state analysis, and simplified design procedures for the ZVZCS PWM full-bridge converter were presented. A 5.4 kW prototype provided experimental results to verify the theoretical analysis.

References

1. Cho, J.G., Sabate, J.A., Hua, G. and Lee, F.C. (1994) Zero-voltage and zero-current-switching full-bridge PWM converter for high power applications. Proceedings of the IEEE Power Electronics Specialists Conference, pp. 102–108.
2. Cho, J.G., Rim, G.H. and Lee, F.C. (1996) Zero voltage and zero current switching full bridge PWM converter with secondary active clamp. Proceedings of the IEEE Power Electronics Specialists Conference, pp. 657–663.
3. Cho, J.G., Baek, J.W., Jeong, C.Y. *et al.* (1998) Novel zero-voltage and zero-current-switching (ZVZCS) full bridge PWM converter using a simple auxiliary circuit. Proceedings of the IEEE Applied Power Electronics Conference and Exposition, pp. 834–839.
4. Kim, E.S., Joe, K.Y., Kye, M.H. *et al.* (1998) An improved ZVZCS PWM full-bridge dc/dc converter using energy recovery snubber. Proceedings of the IEEE Applied Power Electronics Conference and Exposition, pp. 1014–1019.
5. Cho, J.G., Baek, J.W., Yoo, D.W. and Lee, H.S. (1997) Novel zero-voltage and zero-current-switching (ZVZCS) full bridge PWM converter using transformer auxiliary winding. Proceedings of the IEEE Power Electronics Specialists Conference, pp. 227–232.
6. Ruan, X. and Yan, Y. (2001) A novel zero-voltage and zero-current-switching PWM full bridge converters using two diodes in series with the lagging leg. *IEEE Trans. Ind. Electron.*, **48** (4), 777–785.

6

Zero-Voltage-Switching PWM Full-Bridge Converters with Clamping Diodes

6.1 Introduction

In the analysis of Chapters 2–5, the output rectifier diode of the full-bridge converter was taken as an ideal device and its junction capacitor was ignored. In practice, the junction capacitor of the output rectifier diode resonates with the leakage inductor of the transformer or the additional resonant inductor, leading to voltage oscillation and voltage spike across the output rectifier diode. Resistor–capacitor (RC) or resistor–capacitor–diode (RCD) snubber circuits are usually used to damp the voltage oscillation in order to reduce the voltage stress of the output rectifier diode. However, the voltage oscillation cannot be completely eliminated by the RC or RCD snubber circuits and the voltage spike may still be large. Moreover, the conversion efficiency is degraded due to the excessive resonance energy stored in the resonant inductor being dissipated on the resistors of the snubber circuits [1, 2]. In order to completely eliminate the voltage oscillation and voltage spike across the output rectifier diode, two diodes have been introduced at the junction of the resonant inductor and transformer primary winding [3–5]. The two ends of the diodes are connected to the input voltage positive and negative terminals, respectively. The two introduced diodes are used to clamp the transformer primary voltage at V_{in} or $-V_{in}$, where V_{in} is the input voltage. Thus, the transformer secondary voltages is clamped at the reflected input voltage to the secondary side, eliminating the voltage oscillation across the output rectifier diode. The two diodes are referred to as clamping diodes. In this approach, the transformer is connected to the leading leg and the clamping diode conducts twice in a switching period. However, only one conduction of the clamping diode is used for voltage clamp, thus eliminating the voltage oscillation across the output rectifier diodes. The clamping diode will only conduct once in a switching period if the

Soft-Switching PWM Full-Bridge Converters: Topologies, Control, and Design, First Edition. Xinbo Ruan.

positions of the transformer and resonant inductor are exchanged. At this time, the transformer is connected to the lagging leg. Thus, the conduction of the clamping diode, which is unrelated to the voltage clamping, disappears, and the current stress of the clamping diode is reduced [6]. Furthermore, the transformer primary current and the resonant inductor current in the zero state are reduced, leading to reduced conduction losses, improved conversion efficiency, and reduced duty cycle loss.

This chapter analyzes the cause of voltage oscillation and voltage spike across the output rectifier diode in detail and introduces several suppression methods. In particular, the basic idea of eliminating the voltage oscillation and voltage spike with the clamping diodes is explained. Furthermore, the operating principles of the zero-voltage-switching (ZVS) pulse-width-modulation (PWM) full-bridge converter with clamping diodes is analyzed in detail, comparing the situation when the transformer is connected to the leading versus the lagging leg. Experimental measurements from a 3 kW prototype are presented for illustration and verification.

6.2 Causes of Voltage Oscillation in the Output Rectifier Diode in ZVS PWM Full-Bridge Converters

Figure 6.1a shows a ZVS PWM full-bridge converter with phase-shift control, in which Q_1 and Q_3 form the leading leg and Q_2 and Q_4 the lagging leg. The difference between Figure 6.1a and the circuits in Chapter 3 is that junction capacitors C_{DR1} and C_{DR2} are added to the former to simulate the reverse recovery of the output rectifier diodes. Figure 6.1b shows the key waveforms of the full-bridge converter, taking into account the output rectifier diode junction capacitors. Prior to t_5, Q_2 and Q_3 conduct and primary current i_p increases linearly in the negative direction. This current is however insufficient to provide the load current, and hence both D_{R1} and D_{R2} conduct simultaneously. At t_5, i_p reaches the reflected output filter inductor current (i.e., $-i_{Lf}/K$, where K is the primary-to-secondary-windings-turns ratio of the transformer) and D_{R1} turns off. After this, L_r resonates with C_{DR1} and i_p continues to increase in the negative direction while the junction capacitor of D_{R1}, C_{DR1}, is charged, as shown in Figure 6.2. The equivalent circuit is shown in Figure 6.3, where L_f, C_f, and R_{Ld} are replaced by a current source i_{Lf}, i_{Lf}/K is reflected i_{Lf} to the primary side, C'_{DR1} is the reflected C_{DR1} to the primary side, and $C'_{DR1} = 4C_{DR1}/K^2$. It should be noted that L_f is quite large and hardly participates in the resonance between L_r and C_{DR1}. Thus, i_{Lf} rises linearly with the rate of $(V_{in}/K - V_o)/L_f$, as explained in Chapter 3.

At t_5, $I_p(t_5) = -i_{Lf}/K$ and $V_{C'_{DR1}}(t_5) = 0$, initially. Referring to Figure 6.3, i_p and the voltage of C'_{DR1} can be expressed as:

$$i_p(t) = -\left(\frac{i_{Lf}(t)}{K} + \frac{V_{in}}{Z_r}\sin\omega_r(t - t_5)\right) \tag{6.1}$$

$$v_{C'_{DR1}}(t) = V_{in}[1 - \cos\omega_r(t - t_5)] \tag{6.2}$$

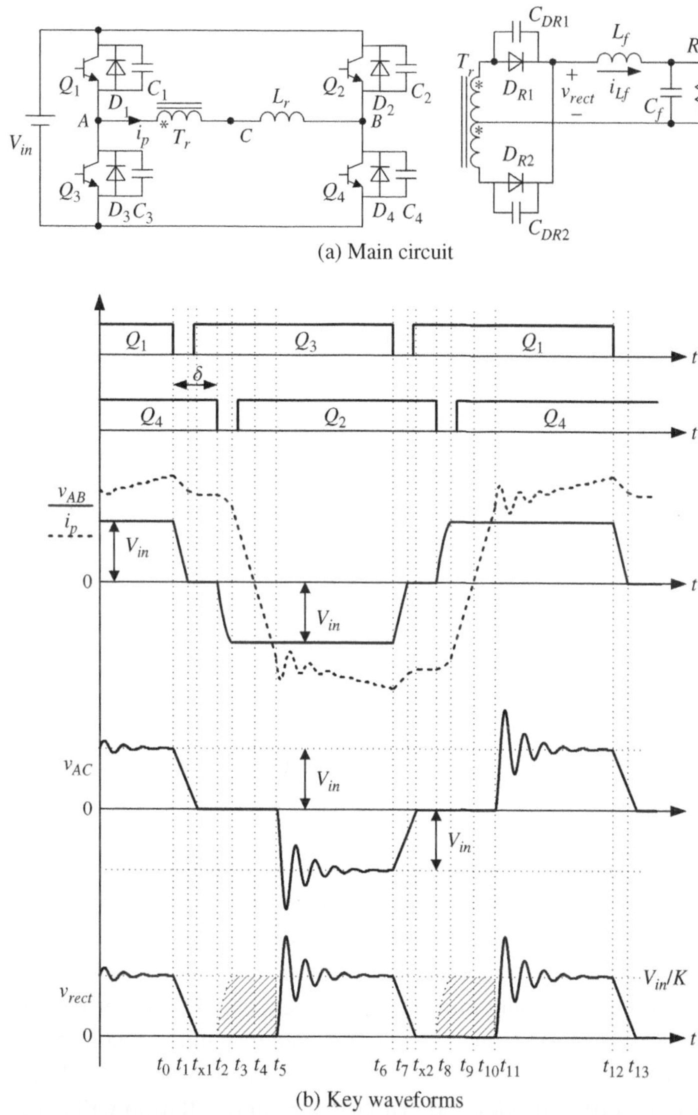

Figure 6.1 A ZVS PWM full-bridge converter, taking into account the junction capacitors of the output rectifier diode

where $Z_r = \sqrt{L_r/C'_{DR1}}$ and $\omega_r = 1/\sqrt{L_r C'_{DR1}}$. Also, v_{CDR1} and v_{rect} are:

$$v_{CDR1}(t) = \frac{2V_{in}}{K}[1 - \cos\omega_r(t - t_5)] \tag{6.3}$$

$$v_{rect}(t) = \frac{V_{in}}{K}[1 - \cos\omega_r(t - t_5)] \tag{6.4}$$

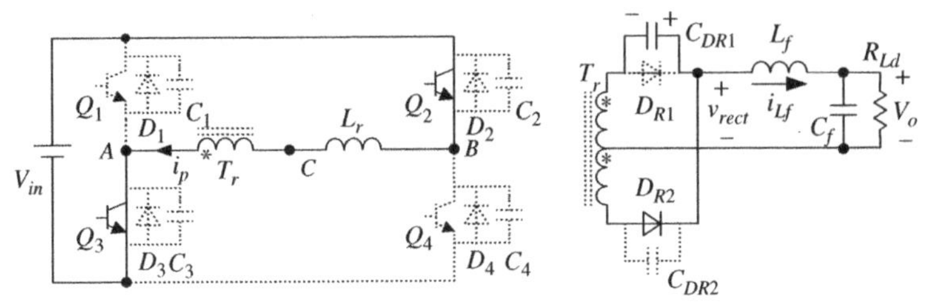

Figure 6.2 L_r resonating with C_{DR1} during $[t_5, t_6]$

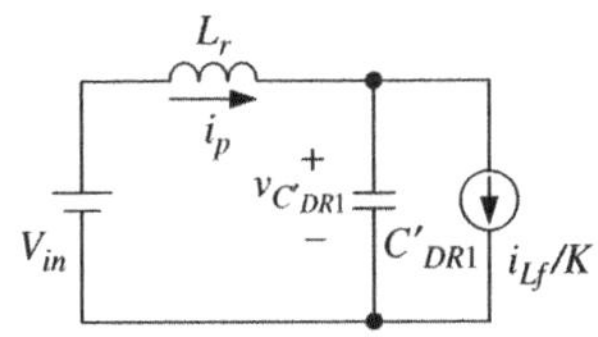

Figure 6.3 Equivalent circuit of Figure 6.2

From Equation 6.1, the maximum value of i_p is:

$$i_p(t) = \frac{i_{Lf}(t)}{K} + \frac{V_{in}}{Z_r} \tag{6.5}$$

The maximum value of i_p includes two parts. The first is the output filter inductor current reflected to the primary side; that is, i_{Lf}/K. The second is V_{in}/Z_r, which is caused by the resonance between the resonant inductor and junction capacitor, and is called the resonant current. In normal operation, i_p is equal to i_{Lf}/K (i.e., the first part of the current) and the resonant current V_{in}/Z_r is excessive; the corresponding energy is stored in the resonant inductor. The excessive energy in the resonant inductor must be dissipated or delivered to input voltage source or load or else voltage oscillation across the output rectifier diodes and primary current oscillation will occur. This phenomenon will be analyzed later.

Equation 6.3 implies that the peak value of the voltage oscillation across the output rectifier diode is $4V_{in}/K$ if the power stage parasitic resistor is zero. This peak value is twice that of the full-bridge converter, assuming the output rectifier diode is an ideal device. In practice, parasitic resistance always exists in the power stage and the voltage oscillation is damped, causing the voltage across the output rectifier diode to converge to $2V_{in}/K$. Also, v_{rect} and i_p converge to V_{in}/K and i_{Lf}/K, respectively, as shown in Figure 6.1b. It can be seen that it is necessary to suppress or even eliminate the voltage oscillation in order to reduce the voltage stress of the output rectifier diode.

6.3 Voltage Oscillation Suppression Approaches

There are many approaches to the suppression or elimination of voltage oscillation, which are generalized into the following five categories:

1. RC snubber;
2. RCD snubber;
3. active clamp circuit [7];
4. auxiliary winding of transformer and clamping diode circuit;
5. clamping diode circuit.

6.3.1 RC Snubber

The use of an RC snubber is the most basic and most common approach to suppressing voltage oscillation, as shown in Figure 6.4. The RC circuit, connected in parallel with the turned-off rectifier diode, suppresses the voltage oscillation when the two output rectifier diodes complete the current commutation, and the excessive energy stored in the resonant inductor is dissipated in the RC circuit's resistor. C_{s1} or C_{s2} will be discharged when the turned-off rectifier diode conducts again, and the energy stored in C_{s1} or C_{s2} will be dissipated in R_{s1} or R_{s2}. Hence, the RC snubber is lossy and the conversion efficiency is degraded.

6.3.2 RCD Snubber

An improved snubber is shown in Figure 6.5 [1, 2], comprising clamping diode D_s, clamping capacitor C_s with larger capacitance, and resistor R_s. The excessive energy stored in the resonant inductor is absorbed by C_s and the voltage oscillation across the output rectifier diode is suppressed. When $v_{AB}=0$, C_s is discharged through R_s until its voltage decays to V_o, with part of the energy dissipated in R_s and part delivered to the load. The loss of the RCD snubber is smaller than that of the RC snubber.

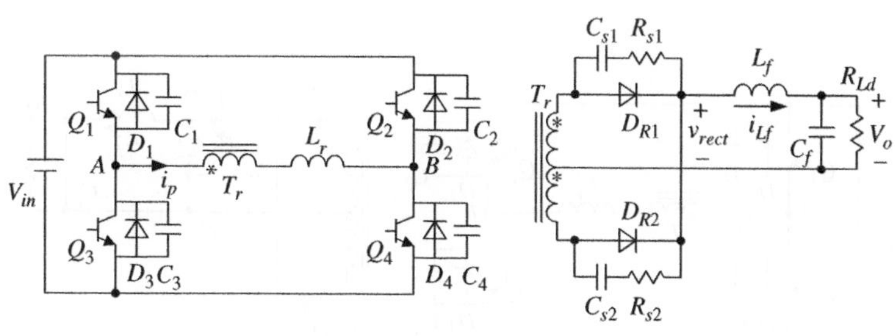

Figure 6.4 RC snubber

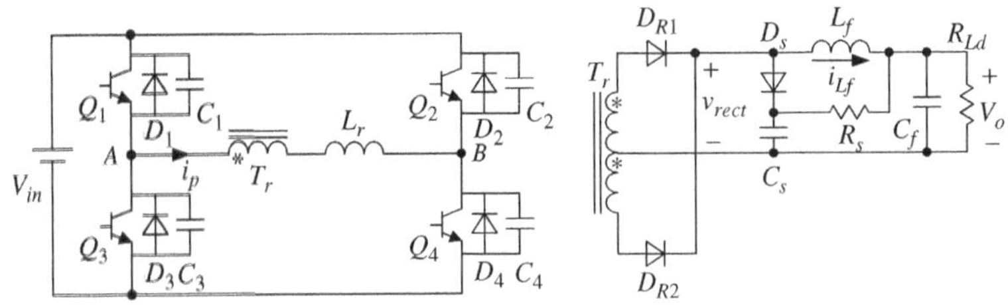

Figure 6.5 RCD snubber

6.3.3 Active Clamp Circuit

Some energy is still dissipated in R_s, although the RCD snubber incurs a smaller amount of loss. The desired method is to deliver all of the excessive energy stored in C_s to the load. So, a switch Q_s is incorporated to be connected in antiparallel with the clamping diode D_s, as shown in Figure 6.6. All of the excessive energy stored in C_s is delivered to the load through Q_s. As a result, resistor R_s is no longer necessary and can be removed. Moreover, Q_s is turned both on and off under zero-voltage conditions and no switching loss is incurred. Therefore, the active clamp circuit not only suppresses the voltage oscillation but also eliminates the loss. However, a switch and corresponding control circuit and driver are required, which increases the cost.

6.3.4 Auxiliary Winding of Transformer and Clamping Diode Circuit

Figure 6.7 shows a voltage oscillation-suppression circuit proposed in reference [8], which includes an auxiliary winding W_c, a buffer capacitor C_c, and clamping diodes $D_{c1} \sim D_{c4}$. The voltage stress of the output rectifier diode is clamped at $2V_{in}/K_c$, where K_c is the turns ratio of auxiliary winding W_c to secondary winding of the transformer. K_c is designed to be slightly smaller than the turns ratio of the transformer K, in order to ensure normal operation of the full-bridge converter. Hence, the voltage across the output rectifier diode is slightly higher than $2V_{in}/K$.

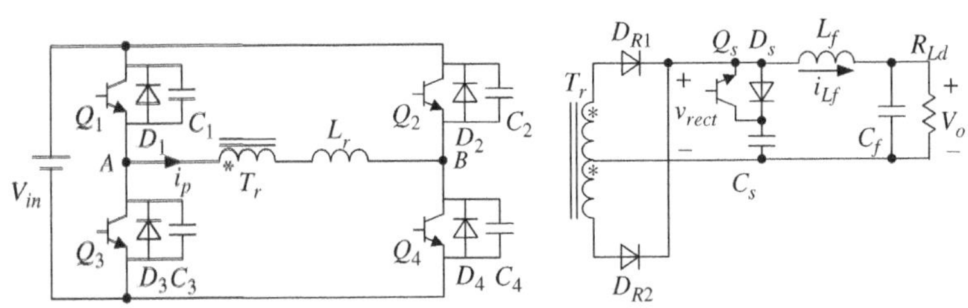

Figure 6.6 Active-clamp circuit

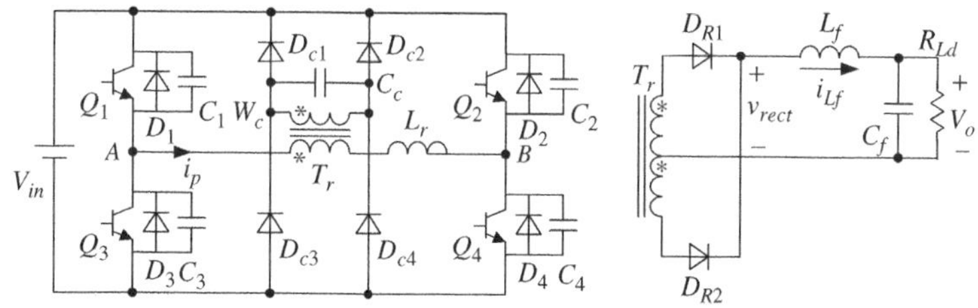

Figure 6.7 Auxiliary winding and clamping diode circuit

6.3.5 Clamping Diode Circuit

In Figures 6.1b and 6.2, Q_2 and Q_3 conduct during $[t_5, t_6]$, and the voltage of node A is 0. The transformer primary voltage v_{AC} is negative. There is thus oscillation with magnitude larger than V_{in}, which means that the voltage of node C may be higher than V_{in}. It will be clamped at V_{in} if a diode D_{c1} is inserted between node C and the positive terminal of the input voltage source. Thus, the negative maximum value of v_{AC} is $-V_{in}$. Accordingly, the maximum voltage across D_{R1} is clamped at $2V_{in}/K$, which is half that without D_{c1}.

Similarly, Q_1 and Q_4 conduct during $[t_{11}, t_{12}]$, and v_{AC} is a positive oscillation waveform with magnitude larger than V_{in}. This means that the voltage of node C may be lower than zero. It will be clamped at zero if a diode D_{c2} is inserted between node C and the negative terminal of the input voltage source. Thus, the positive maximum value of v_{AC} is V_{in} and the maximum voltage across D_{R2} is also clamped at $2V_{in}/K$.

Figure 6.8a shows a ZVS PWM full-bridge converter with D_{c1} and D_{c2}, in which the transformer is connected to the leading leg. The topology is thus defined as

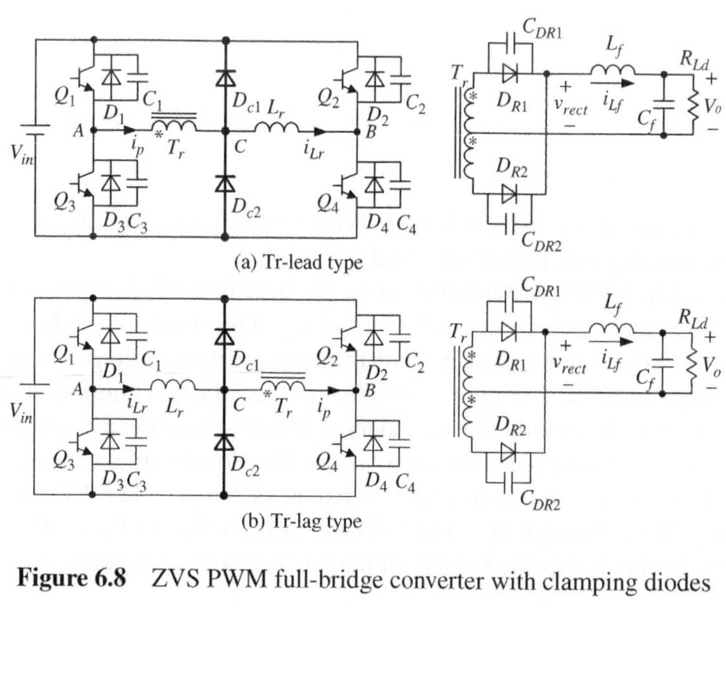

Figure 6.8 ZVS PWM full-bridge converter with clamping diodes

a Tr-lead-type full-bridge converter and D_{c1} and D_{c2} are defined as clamping diodes.

If we exchange the positions of the transformer and the resonant inductor such that the transformer is connected to the lagging leg, as shown in Figure 6.8b, then during $[t_5, t_6]$, v_{CB} is a negative-oscillation waveform with magnitude larger than V_{in}. In this case, the voltage of node B is clamped at V_{in}, due to the conduction of Q_2. It is obvious that the voltage of node C may be lower than zero, and D_{c2} conducts, clamping the voltage of node C at zero. Similarly, during $[t_{11}, t_{12}]$, v_{CB} is a positive oscillation waveform with magnitude larger than V_{in}. In this case, the voltage of node B is clamped at zero due to the conduction of Q_4. It is obvious that the voltage of node C may be higher than V_{in}, and D_{c1} conducts, clamping the voltage of node C at V_{in}. Therefore, the voltage oscillation across the output rectifier diode can also be eliminated by the clamping diodes D_{c1} and D_{c2} when the transformer is connected to the lagging leg. The topology shown in Figure 6.8b is defined as a Tr-lag-type full-bridge converter.

6.4 Operating Principle of the Tr-Lead-Type ZVS PWM Full-Bridge Converter

The key waveforms of the Tr-lead-type ZVS PWM full-bridge converter are shown in Figure 6.9. There are 18 switching modes in a switching cycle. Since the first half-period is similar to the second, only the first is given here for brevity, which includes 9 switching modes and 1 initial switching mode. To simplify the analysis, the following assumptions are made:

1. All the switches and diodes are ideal, except for the output rectifier diode, which is equivalent to an ideal diode and a parallel-connected capacitor.
2. All the capacitors, inductors, and transformer are ideal.
3. $C_1 - C_3 = C_{lead}$, $C_2 = C_4 = C_{lag}$, $C_{DR1} = C_{DR2} = C_{DR}$.
4. $L_r \ll K^2 L_f$, where K is the primary-and-secondary-windings-turns ratio of the transformer.
5. The leakage inductance of the transformer is too small to be ignored.

Figure 6.10 gives the equivalent circuits of each switching mode in a half-cycle.

1. **Mode 0, prior to t_0 (Figure 6.10a):** Prior to t_0, Q_1, Q_4, and D_{R1} are conducting and the primary side powers the load.
2. **Mode 1, $[t_0, t_1]$ (Figure 6.10b):** At t_0, Q_1 is turned off, i_p charges C_1 and discharges C_3, and v_{AB} declines. At the beginning of this mode, v_{AC} is kept unchanged and the voltage of node C decays with the decrease in the voltage of node A. When the voltage of node C decreases to zero, D_{c2} conducts, clamping it at zero. v_{AC} then begins to decay, and the secondary rectifier voltage decays correspondingly. Thus, C_{DR2} is discharged and i_p decreases with a downward moment. As D_{c2} conducts, L_r is shorted and its current i_{Lr} remains unchanged. The difference between i_p and i_{Lr} flows through D_{c2}. The fall time of the voltage of node C is very short and can be ignored as the voltage of node C is nearly zero before t_0.

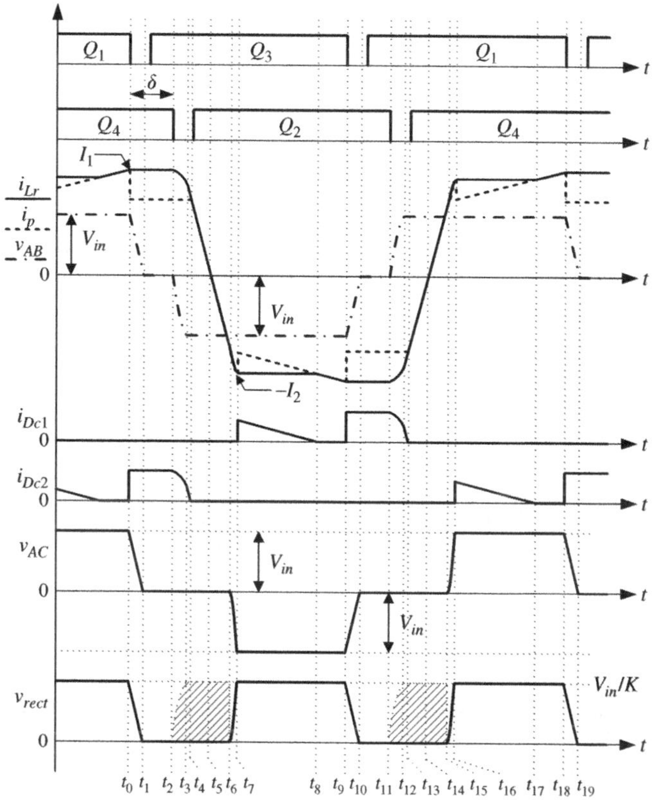

Figure 6.9 Key waveforms of the Tr-lead-type ZVS PWM full-bridge converter

This mode can be further simplified as shown in Figure 6.11, where C'_{DR2} is the reflected C_{DR2} to the primary side. It can be seen from Figure 6.11 that a part of i_{Lf} discharges C_{DR2}, while the rest is reflected to the primary side to charge C_1 and discharge C_3. Thus, v_{C1}, v_{C3}, i_p, and i_{Lr} can be expressed as:

$$v_{C1}(t) = \frac{I_1}{2C_{lead} + C'_{DR}}(t - t_0) \tag{6.6}$$

$$v_{C3}(t) = v_{C'_{DR2}}(t) = V_{in} - \frac{I_1}{2C_{lead} + C'_{DR}}(t - t_0) \tag{6.7}$$

$$i_{Lr}(t) = I_1 \tag{6.8}$$

$$i_p(t) = \frac{2C_{lead}}{2C_{lead} + C'_{DR}} I_1 \tag{6.9}$$

where I_1 is the output filter inductor current reflected to the primary side at t_0 and $C'_{DR} = 4C_{DR}/K^2$.

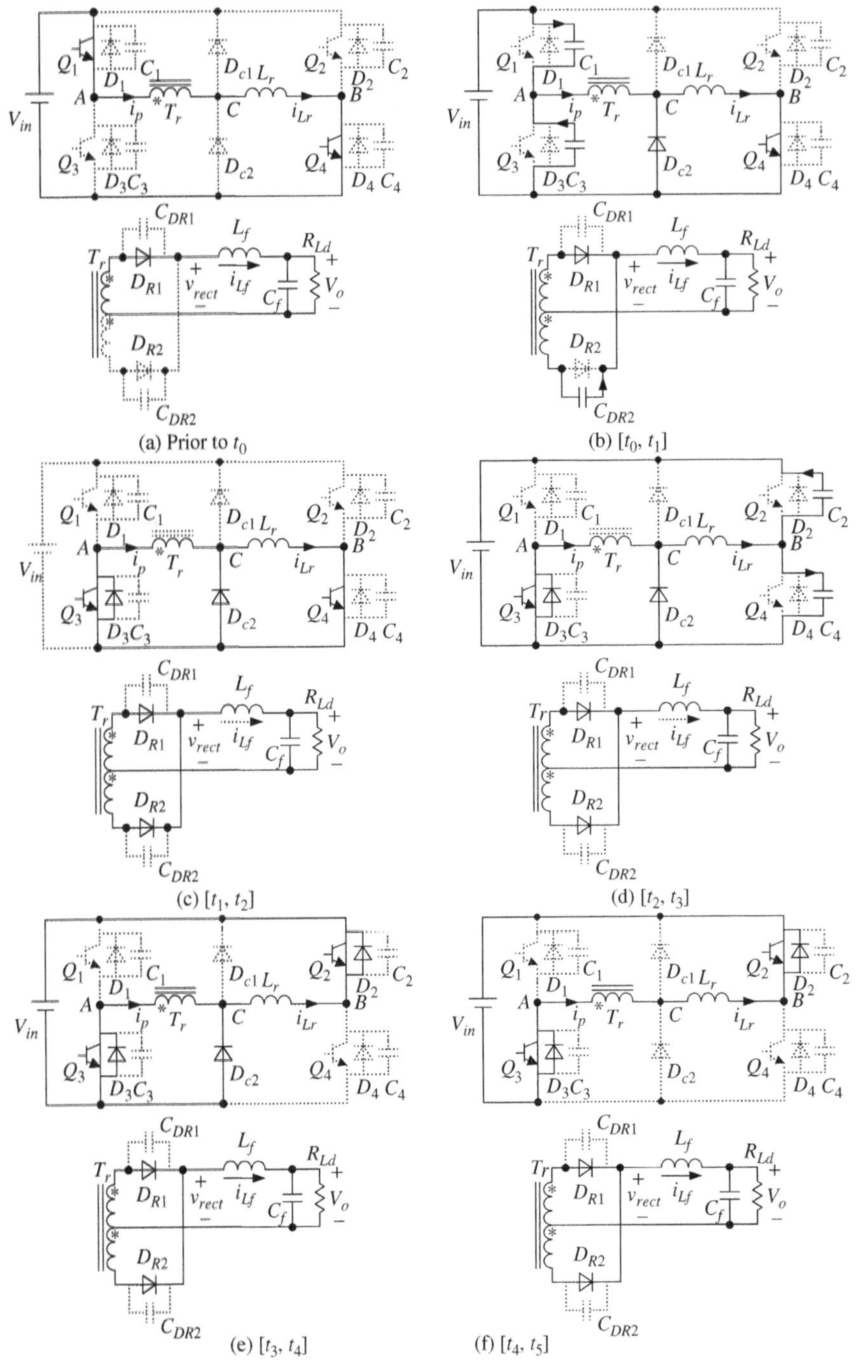

Figure 6.10 Equivalent circuits of the individual modes of the Tr-lead-type ZVS PWM full-bridge converter

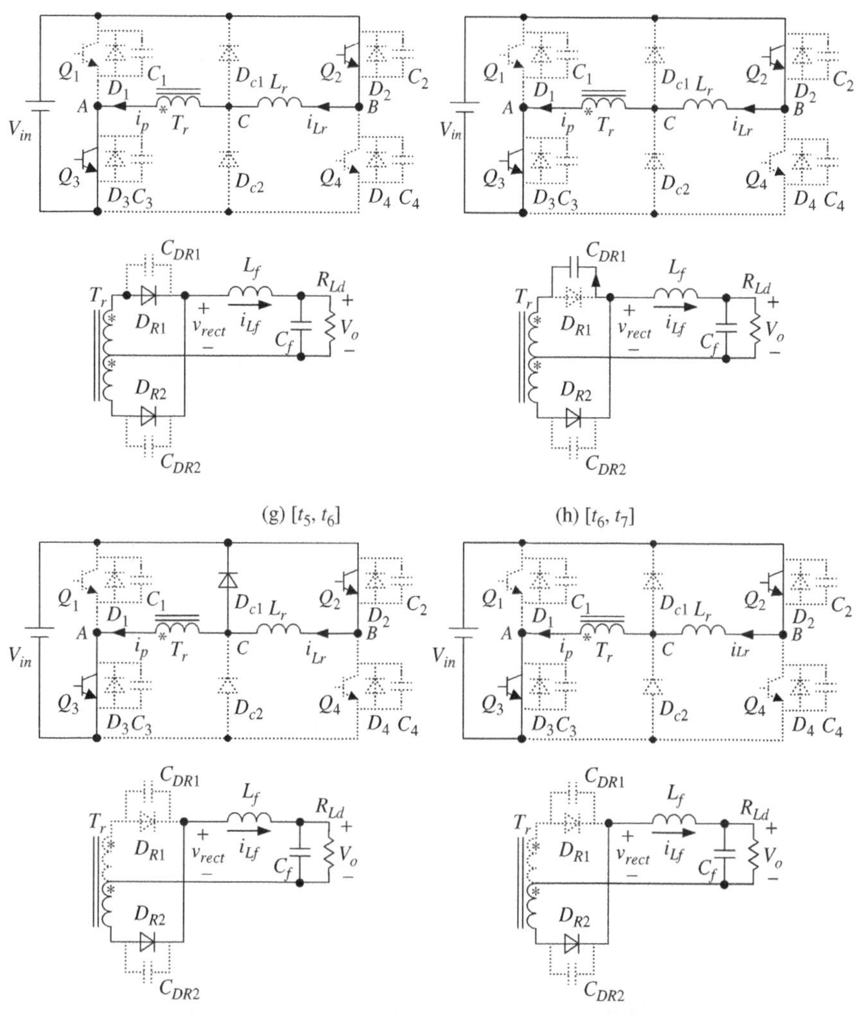

(g) $[t_5, t_6]$ (h) $[t_6, t_7]$

(i) $[t_7, t_8]$ (j) $[t_8, t_9]$

Figure 6.10 *(Continued)*

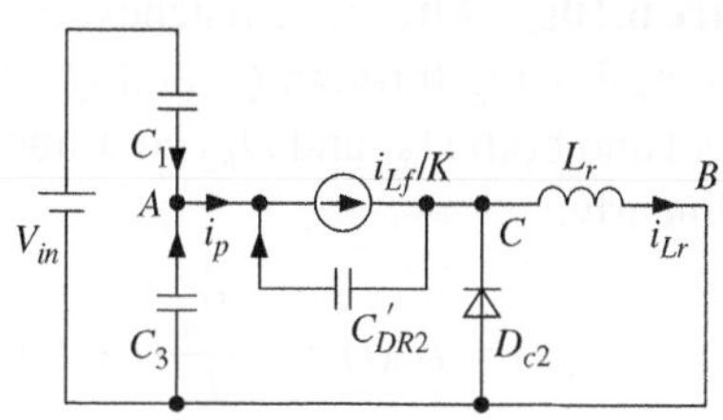

Figure 6.11 Simplified equivalent circuit of mode 1

Equation 6.9 indicates that i_p decreases with a downward moment at t_0. Meanwhile, Q_1 is turned off under zero-voltage conditions thanks to C_1, C_3, and C_{DR2}.

3. **Mode 2, [t_1, t_2] (Figure 6.10c):** At t_1, v_{C1} reaches V_{in}, while v_{C3} and v_{CDR2} decay to zero. Thus, D_{R2} and the antiparallel diode of Q_3 conduct naturally and Q_3 can be turned on with zero voltage. In this mode, $v_{AB}=0$, i_p and i_{Lr} are kept unchanged and D_{R1} and D_{R2} conduct simultaneously.
4. **Mode 3, [t_2, t_3] (Figure 6.10d):** At t_2, Q_4 is turned off with zero voltage thanks to C_2 and C_4, while i_{Lr} charges C_4 and discharges C_2. Moreover, $v_{AB}=-v_{C4}$. As both the output rectifier diodes are conducting, both the primary and secondary voltages of the transformer are zero and $v_{rect}=0$. In this case, v_{AB} is fully applied on L_r, which resonates with C_2 and C_4. Thus, i_{Lr}, v_{C2}, and v_{C4} are expressed as:

$$i_{Lr}(t) = I_1 \cos \omega_1 (t - t_2) \tag{6.10}$$

$$v_{C4}(t) = Z_1 I_1 \sin \omega_1 (t - t_2) \tag{6.11}$$

$$v_{C2}(t) = V_{in} - Z_1 I_1 \sin \omega_1 (t - t_2) \tag{6.12}$$

where $Z_1 = \sqrt{L_r/(2C_{lag})}$ and $\omega_1 = 1/\sqrt{2L_r C_{lag}}$.

This mode ends at t_3 when v_{C4} reaches V_{in} and v_{C2} decays to 0.

5. **Mode 4, [t_3, t_4] (Figure 6.10e):** At t_3, D_2 conducts naturally and Q_2 can be turned on with zero voltage. In this mode, since both D_{R1} and D_{R2} still conduct simultaneously, both the primary and the secondary voltages of the transformer are zero and $v_{rect}=0$. D_{c2} then continues to conduct and V_{in} is fully applied on L_r, causing its current i_{Lr} to decay linearly:

$$i_{Lr}(t) = I_{Lr}(t_3) - \frac{V_{in}}{L_r}(t - t_3) \tag{6.13}$$

At t_4, i_{Lr} decays to i_p and D_{c2} turns off naturally.

6. **Mode 5, [t_4, t_5] (Figure 6.10f):** In this mode, $v_{rect}=0$ and $v_{AC}=0$. V_{in} is fully applied on L_r, causing i_{Lr} and i_p to decay linearly:

$$i_p(t) = i_{Lr}(t) = I_{Lr}(t_4) - \frac{V_{in}}{L_r}(t - t_4) \tag{6.14}$$

At t_5, i_p decays to zero and D_2 and D_3 turn off naturally.

7. **Mode 6, [t_5, t_6] (Figure 6.10g):** After t_5, i_p reaches zero and continues to increase in the negative direction, flowing through Q_2 and Q_3. Here, i_p is not sufficient to provide the load current and both D_{R1} and D_{R2} continue to conduct. Thus, $v_{rect}=0$ and i_{Lr} and i_p decay linearly:

$$i_p(t) = i_{Lr}(t) = -\frac{V_{in}}{L_r}(t - t_5) \tag{6.15}$$

At t_6, i_p reaches the reflected filter inductor current $-I_{Lf}(t_6)/K$. At the same time, D_{R1} turns off and D_{R2} carries all of the load current.

8. **Mode 7, [t_6, t_7] (Figure 6.10h):** At t_6, L_r resonates with C_{DR1} and C_{DR1} is charged. At the same time, i_p and i_{Lr} continue to increase. i_p, i_{Lr}, and v_{CDR1} can be expressed as:

$$i_p(t) = i_{Lr}(t) = \frac{I_{Lf}(t_6)}{K} + \frac{V_{in}}{Z_r}\sin\omega_r(t - t_6) \tag{6.16}$$

$$v_{CDR1}(t) = \frac{2V_{in}}{K}[1 - \cos\omega_r(t - t_6)] \tag{6.17}$$

where $Z_r = \sqrt{L_r/C'_{DR}}$ and $\omega_r = 1/\sqrt{L_rC'_{DR}}$.

At t_7, the voltage of C_{DR1} rises to $2V_{in}/K$ and v_{AC} decays to $-V_{in}$. In this case, the voltage of node C rises to V_{in}. Thus, D_{c1} conducts and the voltage of C_{DR1} is clamped at $2V_{in}/K$. i_p and i_{Lr} are $-I_2$, where the value of I_2 is:

$$I_2 = \frac{I_{Lf}(t_6)}{K} + \frac{V_{in}}{Z_r} \tag{6.18}$$

9. **Mode 8, [t_7, t_8] (Figure 6.10i):** When D_{c1} conducts, i_p decays downward to the reflected i_{Lf} (i.e., $i_p = -i_{Lf}/K$), while i_{Lr} is kept unchanged at $-I_2$. The difference between i_{Lr} and i_p flows through D_{c1}. In this mode, i_{Lf} increases linearly, causing i_p to increase linearly in the negative direction, expressed as:

$$i_p(t) = -\frac{i_{Lf}(t)}{K} = -\frac{V_{in} - KV_o}{K^2L_f}(t - t_7) \tag{6.19}$$

The current flowing through D_{c1} decays accordingly. At t_8, i_p is equal to i_{Lr} and D_{c1} turns off naturally.

10. **Mode 9, [t_8, t_9] (Figure 6.10j):** In this mode, the load is powered by the input voltage source, with i_p equal to i_{Lr} (and given by Equation 6.19).

6.5 Operating Principle of the Tr-Lag-Type ZVS PWM Full-Bridge Converter

The operating principle of the Tr-lag-type ZVS PWM full-bridge converter is different from that of the Tr-lead-type. The key waveforms of the former are shown in Figure 6.12.

1. **Mode 0, prior to t_0 (Figure 6.13a):** Prior to t_0, Q_1, Q_4, and D_{R1} conduct and the power is transferred from the input source V_{in} to the load.
2. **Mode 1, [t_0, t_1] (Figure 6.13b):** At t_0, Q_1 is turned off, i_p charges C_1 and discharges C_3, and v_{AB} decays. Here, Q_1 realizes zero-voltage turn-off thanks to C_1 and C_3. v_{CB} and the secondary voltage of the transformer decay simultaneously, discharging C_{DR2}. This mode can be further simplified as shown in Figure 6.14a.

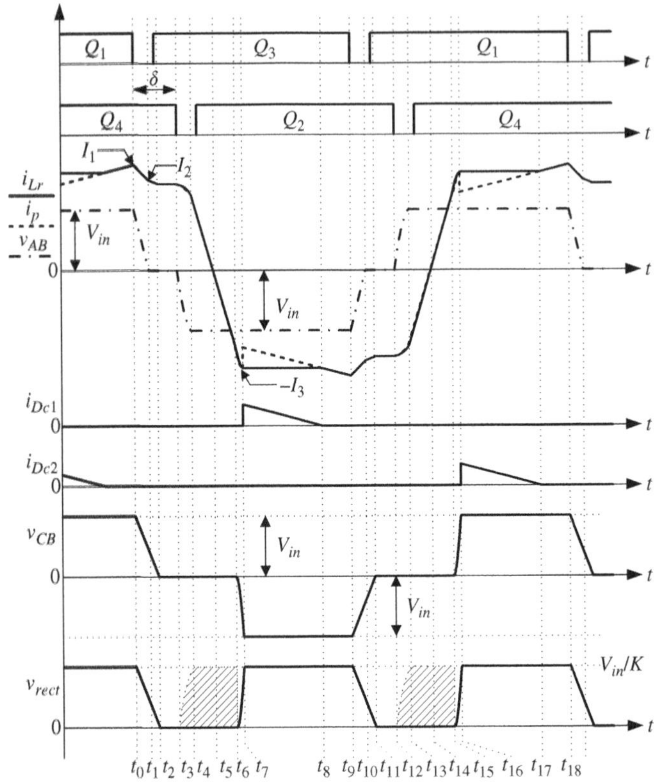

Figure 6.12 Key waveforms of the Tr-lag-type ZVS PWM full-bridge converter

v_{C1}, v_{C3}, $v_{C'_{DR2}}$, i_p, and i_{Lr} are expressed as:

$$v_{C1}(t) = \frac{C'_{DR}}{2\omega_2 C_{lead}(2C_{lead} + C'_{DR})} I_1 \sin\omega_2(t - t_0) + \frac{1}{2C_{lead} + C'_{DR}} I_1(t - t_0) \tag{6.20}$$

$$v_{C3}(t) = V_{in} - \frac{C'_{DR}}{2\omega_2 C_{lead}(2C_{lead} + C'_{DR})} I_1 \sin\omega_2(t - t_0) - \frac{1}{2C_{lead} + C'_{DR}} I_1(t - t_0) \tag{6.21}$$

$$v_{C'_{DR2}}(t) = V_{in} + \frac{1}{\omega_2(2C_{lead} + C'_{DR})} I_1 \sin\omega_2(t - t_0) - \frac{1}{2C_{lead} + C'_{DR}} I_1(t - t_0) \tag{6.22}$$

$$i_p(t) = i_{Lr}(t) = \frac{C'_{DR}}{2C_{lead} + C'_{DR}} I_1 \cos\omega_2(t - t_0) + \frac{2C_{lead}}{2C_{lead} + C'_{DR}} I_1 \tag{6.23}$$

where I_1 is the reflected filter inductor current at t_0 and $\omega_2 = \sqrt{\frac{2C_{lead}+C'_{DR}}{2L_r C_{lead} C'_{DR}}}$.

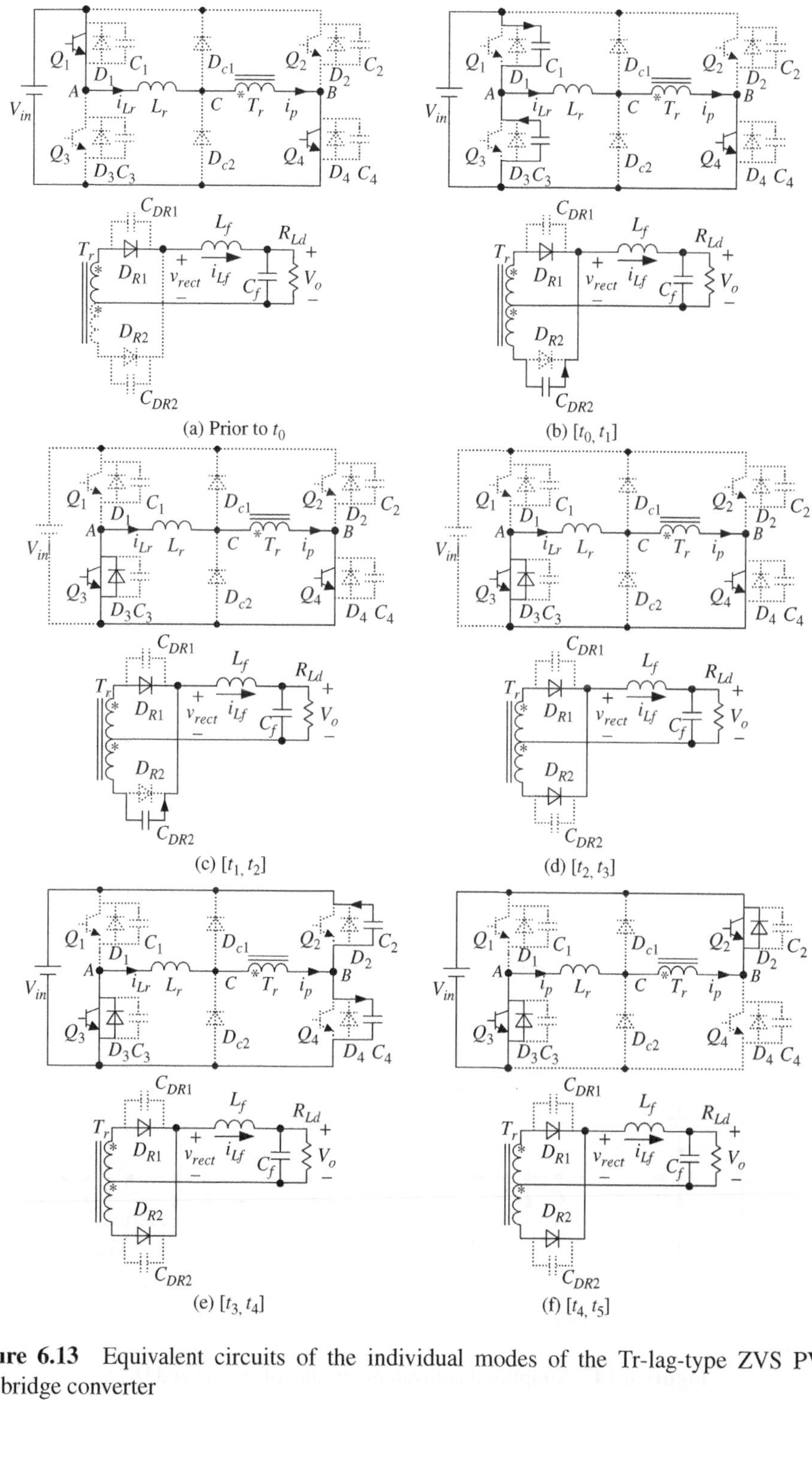

Figure 6.13 Equivalent circuits of the individual modes of the Tr-lag-type ZVS PWM full-bridge converter

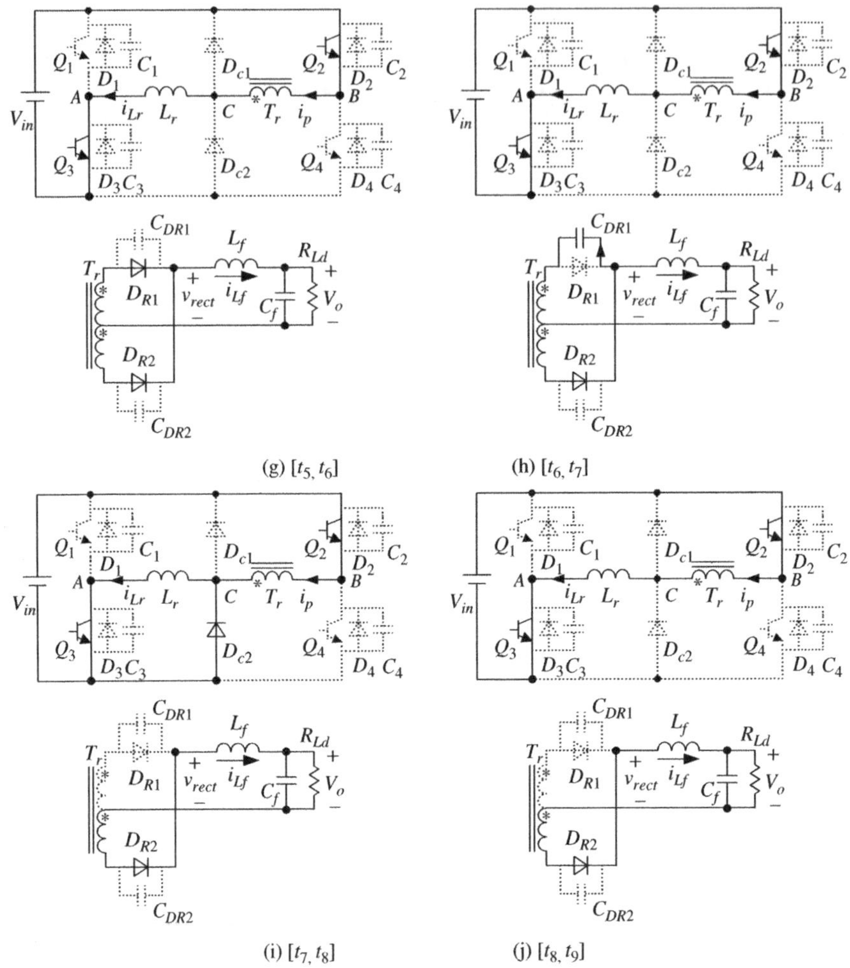

Figure 6.13 (*Continued*)

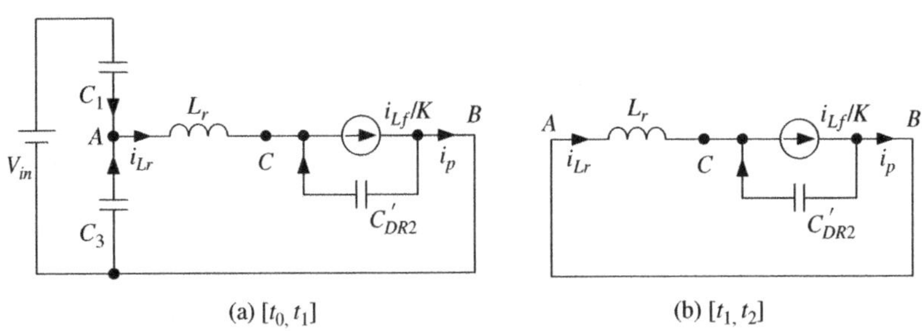

Figure 6.14 Simplified equivalent circuits of [t_0, t_1] and [t_1, t_2]

Due to the very short time interval of $[t_0, t_1]$, v_{C1} and v_{C3} can be approximated as:

$$v_{C1}(t) = \frac{I_1}{2C_{lead} + C'_{DR}}(t - t_0) \tag{6.24}$$

$$v_{C3}(t) = V_{in} - \frac{I_1}{2C_{lead} + C'_{DR}}(t - t_0) \tag{6.25}$$

At t_1, v_{C1} rises to V_{in} and v_{C3} decays to zero. Thus, D_3 conducts.

3. **Mode 2, $[t_1, t_2]$ (Figure 6.13c):** After D_3 conducts, Q_3 can be turned on with zero voltage. The voltage of node C is positive when that of node A decays to zero. At this time, C_{DR2} continues to be discharged and i_{Lr} and i_p continue to decay. The equivalent circuit is further simplified as shown in Figure 6.14b. The expressions of $v_{C'_{DR2}}$, i_p, and i_{Lr} are expressed as:

$$v_{C'_{DR2}}(t) = \frac{1}{\omega_r C'_{DR}}(I_2 - I_1)\sin\omega_r(t - t_1) + V_{C'_{DR2}}(t_1)\cos\omega_r(t - t_1) \tag{6.26}$$

$$i_p(t) = i_{Lr}(t) = (I_2 - I_1)\cos\omega_r(t - t_1) - \frac{V_{C'_{DR2}}(t_1)}{\omega_r L_r}\sin\omega_r(t - t_1) + I_1 \tag{6.27}$$

where I_2 is the primary current at t_1.

At t_2, C_{DR2} is fully discharged and D_{R2} conducts. Thus, the voltage of node C decays to zero.

4. **Mode 3, $[t_2, t_3]$ (Figure 6.13d):** D_{R1} and D_{R2} conduct simultaneously, clamping the primary and secondary voltages of the transformer at zero. The voltages of nodes A, B, and C are zero. i_{Lr} and i_p are equal and freewheeling.
5. **Mode 4, $[t_3, t_4]$ (Figure 6.13e):** At t_3, Q_4 is turned off. Thus, i_{Lr} charges C_4 and discharges C_2. Here, Q_4 achieves zero-voltage turn-off thanks to C_2 and C_4. As both of the rectifier diodes are conducting, both the primary and the secondary voltages of the transformer are zero. v_{AB} is fully applied on L_r. During this interval, L_r resonates with C_2 and C_4. The formulas of v_{C2}, v_{C4}, and i_{Lr} are the same as Equations 6.10–6.12, but with I_1 is replaced by I_2.

   At t_4, the voltage of C_4 rises to V_{in} and that of C_2 decays to zero. Thus, D_2 conducts.
6. **Mode 5, $[t_4, t_5]$ (Figure 6.13f):** In this mode, both D_{R1} and D_{R2} continue conducting simultaneously, with $v_{rect} = 0$ and $v_{CB} = 0$. V_{in} is fully applied on L_r. Therefore, i_{Lr} and i_p decay linearly. At t_5, i_{Lr} and i_p decay to zero, with D_2 and D_3 turned off naturally.
7. **Mode 6, $[t_5, t_6]$ (Figure 6.13g):** After t_5, i_p and i_{Lr} reach zero and continue to increase in the negative direction, flowing through Q_2 and Q_3. Here, i_p is not enough to provide the load current, causing both of the rectifier diodes to continue conducting and v_{rect} to equal zero. Thus, V_{in} is fully applied on L_r, with i_{Lr} and i_p linearly increased in the negative direction. At t_6, i_p reaches the reflected output

filter inductor current $-I_{Lf}(t_6)$. Therefore, D_{R1} turns off and D_{R2} carries all of the load current.

8. **Mode 7, $[t_6, t_7]$ (Figure 6.13h):** From t_6, L_r resonates with C_{DR1}, which is charged. i_p and i_{Lr} thus continue to increase in the negative direction. At t_7, the voltage of C_{DR1} rises to $2V_{in}/K$ and v_{CB} decays to $-V_{in}$. Thus, the voltage of node C reduces to zero, causing D_{c2} to conduct and clamping v_{CB} at $-V_{in}$. The voltage of C_{DR1} is therefore clamped at $2V_{in}/K$, with i_p equal to i_{Lr}. Here, the values of i_p and i_{Lr} are $-I_3$, where the value of I_3 is:

$$I_3 = \frac{I_{Lf}(t_6)}{K} + \frac{V_{in}}{Z_r} \tag{6.28}$$

9. **Mode 8, $[t_7, t_8]$ (Figure 6.13i):** When D_{c2} conducts, i_p decays downward to the reflected i_{Lf} (i.e., $i_p = -i_{Lf}/K$), while i_{Lr} remains unchanged at $-I_3$. The difference between i_{Lr} and i_p flows through D_{c2}. In this mode, i_{Lf} increases linearly in a positive direction and i_p in a negative one. Thus, the current flowing through D_{c2} decays linearly. At t_8, i_p is equal to i_{Lr} and D_{c2} turns off naturally.
10. **Mode 9, $[t_8, t_9]$ (Figure 6.13j):** In this mode, the load is powered by the input voltage source, with i_p equal to i_{Lr} and having the same formula as given in Equation 6.19.

6.6 Comparisons of Tr-Lead-Type and Tr-Lag-Type ZVS PWM Full-Bridge Converters

6.6.1 *Clamping Diode Conduction Times*

As shown in Figure 6.9, the clamping diode conducts twice in a switching cycle when the transformer is connected to the leading leg. In this case, D_{c1} conducts during $[t_7, t_8]$ and $[t_9, t_{13}]$. Only the conduction of $[t_7, t_8]$ is related to the elimination of the voltage oscillation across the output rectifier diode. As shown in Figure 6.12, the clamping diode conducts only once in a switching cycle when the transformer is connected to the lagging leg, and in this case, D_{c1} only conducts during $[t_7, t_8]$ and eliminates the voltage oscillation, and is reverse-biased during $[t_9, t_{13}]$.

When the transformer is connected to the leading leg, the conduction of the clamping diode, which is unrelated to the voltage oscillation elimination, occurs at the transition of the leading-leg switches. It can be seen from Figure 6.10a,b that the voltage of node C is nearly zero because the resonant inductance is much smaller than the reflected filter inductance when Q_1 and Q_4 conduct. The voltage of node A decays when Q_1 is turned off, with v_{AC} decaying simultaneously. Then the voltage of node C also decays and will become zero, forcing D_{c2} to conduct. Similarly, D_{c1} will be forced to conduct when Q_1 is turned off.

When the transformer is connected to the lagging leg, as shown in Figure 6.13b, the voltage of node C is slightly lower than V_{in} before Q_1 is turned off. The voltage of node A decays when Q_1 is turned off, and v_{CB} decays accordingly. The voltage of node B is zero due to the conduction of Q_4. The voltage of node C is greater than zero, although it decays. Therefore, D_{c2} does not conduct. Similarly, D_{c1} does not conduct when Q_3 is turned off.

It can be concluded that, unlike in the Tr-lead-type full-bridge converter, the clamping diode only conducts once in a switching cycle in the Tr-lag-type full-bridge converter. Hence, the current rating of the clamping diode can be reduced.

6.6.2 Achievement of ZVS

6.6.2.1 Leading-Leg Switches

In the Tr-lead-type full-bridge converter shown in Figure 6.11, in order to achieve ZVS for the leading-leg switches, sufficient energy is required to: (i) discharge the junction capacitor of the switch that will be turned on; (ii) charge the junction capacitor of the switch that will be turned off; and (iii) discharge the junction capacitor of one output rectifier diode. All of the energy is provided by the output filter inductor.

In the Tr-lag-type full-bridge converter shown in Figure 6.14a, in order to achieve ZVS for the leading-leg switches, sufficient energy is required to: (i) discharge the junction capacitor of the switch that will be turned on; (ii) charge the junction capacitor of the switch that will be turned off; and (iii) discharge part of the junction capacitor of one output rectifier diode. All of the energy is provided by the output filter inductor and resonant inductor.

Based on this analysis, it can be seen that the leading-leg switches in the Tr-lag-type full-bridge converter achieve ZVS more readily than those of the Tr-lead-type full-bridge converter.

6.6.2.2 Lagging-Leg Switches

In order to achieve ZVS for the lagging-leg switches of both the Tr-lead-type and the Tr-lag-type full-bridge converters, sufficient energy is required to discharge the junction capacitor of the switch that will be turned on and to charge the junction capacitor of the switch that will be turned off. Hence, the required energy is the same for the two converters.

For the Tr-lead-type full-bridge converter, the resonant inductor is shorted by the clamping diode after the leading-leg switch is turned off. The resonant inductor current is then kept unchanged at the reflected filter inductor current until the lagging-leg switch is turned off. For the Tr-lag-type full-bridge converter, the resonant inductor resonates with the junction capacitors of two leading-leg switches and with the incoming output rectifier diode after the leading-leg switch is turned off. The resonant inductor current then begins to decay. Therefore, the resonant inductor current in the Tr-lag-type full-bridge converter is smaller than that in the Tr-lag-type full-bridge converter at the turn-off instance of the lagging-leg switch. Thus, it is more difficult to achieve ZVS in the lagging-leg switches of the Tr-lag-type converter than in those of the Tr-lead-type converter.

6.6.3 *Conduction Loss in Zero State*

As shown in Figures 6.9 and 6.12, the resonant inductor current in the zero state of the Tr-lag-type full-bridge converter is smaller than that of the Tr-lead-type full-bridge converter. Therefore, the conduction loss in the primary side is reduced, resulting in a higher efficiency.

6.6.4 *Duty Cycle Loss*

The duty cycle loss is proportional to the time it takes for the resonant inductor current to transit from the positive (or negative) value to the negative (or positive) reflected output filter inductor current. As the initial transition current of the resonant inductor in the Tr-lag-type converter is smaller than that in the Tr-lead-type converter, the duty cycle loss is reduced. Thus, the turns ratio of the primary-to-secondary-windings of the transformer can be increased to further reduce the conduction loss, and a higher efficiency can be expected.

6.6.5 *Effect of the Blocking Capacitor*

In practice, the conduction time and voltage drop of diagonal switches Q_1 and Q_4 cannot be matched with those of diagonal switches Q_2 and Q_3. In other words, v_{AB} is not a pure ac voltage but contains a dc component. Since the winding resistance of the high-frequency transformer is very small, the transformer will be saturated if the dc component exists for a long time, no matter how small it is. Hence, blocking of the dc component of v_{AB} is an important issue for the full-bridge converter. Peak current control is usually used to block the dc component. The transformer saturation can be avoided by ensuring that the peak currents of Q_1 and Q_4 are the same as those of Q_2 and Q_3. Another method is to sense the dc component of v_{AB}. Specifically, the conduction time of Q_1 and Q_4 (or Q_2 and Q_3) is reduced when the dc component is positive (or negative), eliminating the latter. In practice, a simple way of preventing the transformer from saturating is to insert a blocking capacitor C_b at the primary side in series with the transformer or the resonant inductor. On introducing the blocking capacitor to the Tr-lead-type and Tr-lag-type converters, four converters can be obtained, as shown in Figure 6.15.

6.6.5.1 Tr-Lead-Type Full-Bridge Converter

In the Tr-lead-type full-bridge converter, the clamping diode conducts twice in a switching cycle. When it occurs in the zero state, both the transformer and the resonant

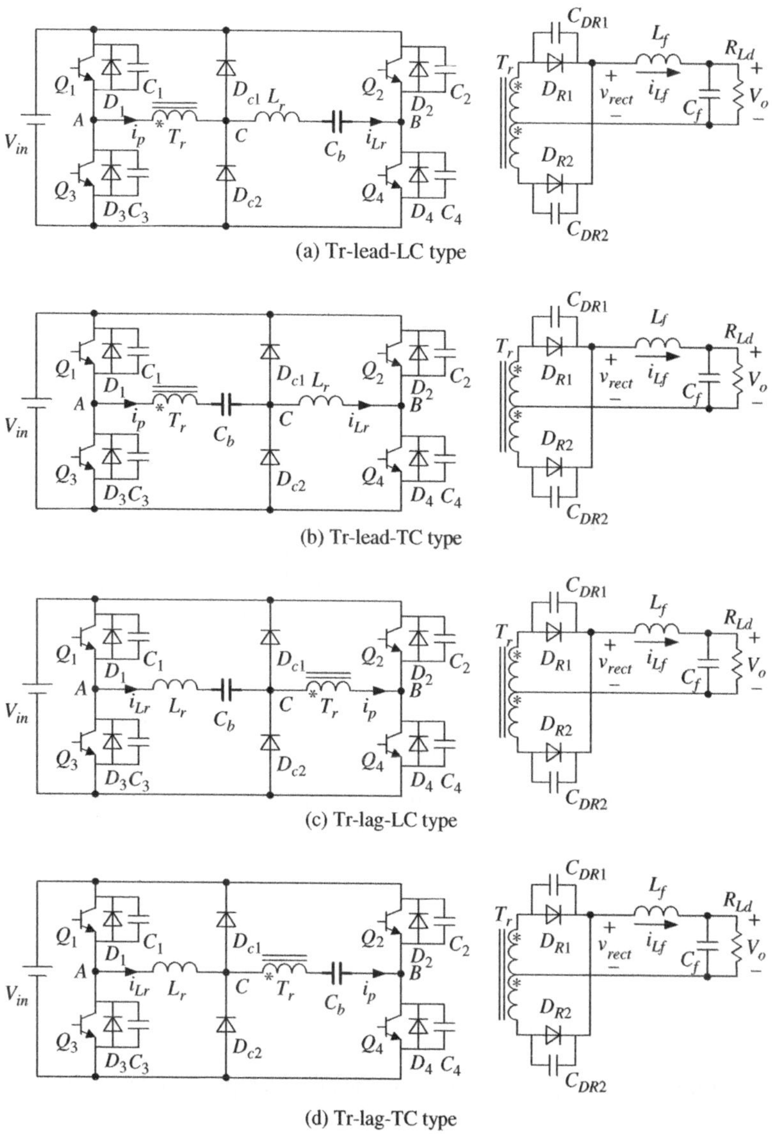

Figure 6.15 Four full-bridge converters with blocking capacitors

inductor are shorted. When it occurs after the output rectified voltage is clamped, only the resonant inductor is shorted. Figure 6.16 shows the schematics when the clamping diode conducts (only the primary side is shown). If C_b is in series with the resonant inductor when the resonant inductor is shorted, its equivalent circuit will be as shown in Figure 6.17a. The dc component of C_b, which is equal to the dc component of v_{AB}, will result in asymmetry of i_L in the positive and negative directions. If C_b is in series with the transformer when the transformer is shorted, its equivalent circuit will be as

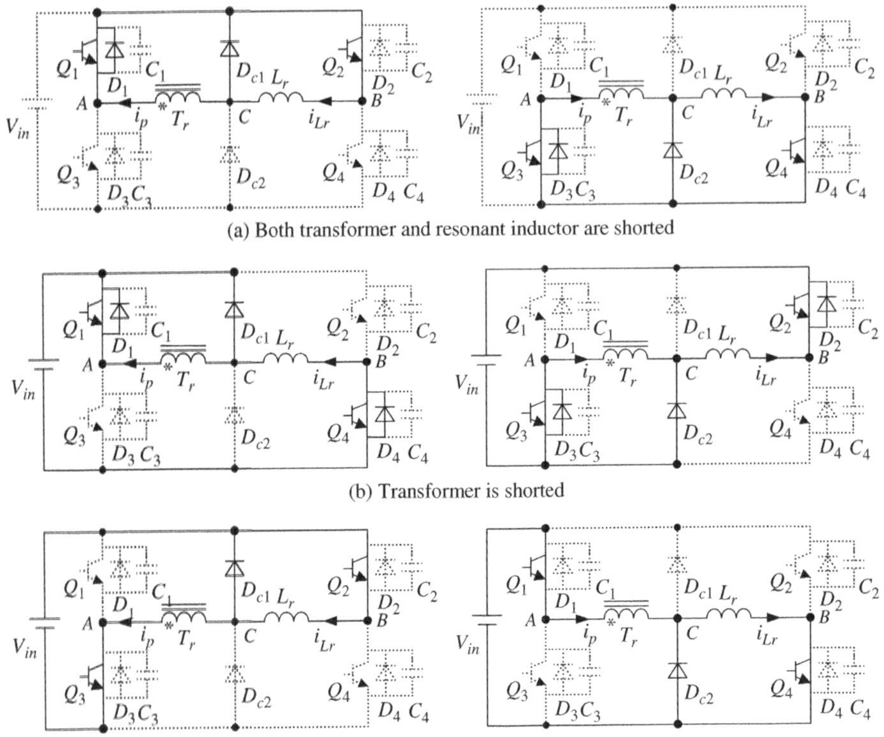

(a) Both transformer and resonant inductor are shorted

(b) Transformer is shorted

(c) Resonant inductor is shored

Figure 6.16 Equivalent circuits when the clamping diode conducts in the Tr-lead-type full-bridge converter (left side: D_{c1} conducts; right side: D_{c2} conducts)

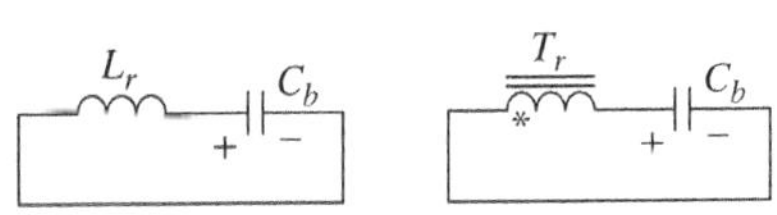

(a) L_r series with C_b (b) T_r series with C_b

Figure 6.17 Equivalent circuits when the clamping diode conducts

shown in Figure 6.17b. In this case, the dc component of v_{AB} will be applied on the leakage inductor and will result in asymmetry of i_p in the positive and negative directions. The asymmetry of i_p or i_{Lr} will result in asymmetry between the currents of two clamping diodes.

6.6.5.2 Tr-Lag-Type Full-Bridge Converter

In the Tr-lag-type full-bridge converter, the clamping diodes only conduct after the output rectified voltage is clamped, as shown in Figure 6.18. At that time, only the

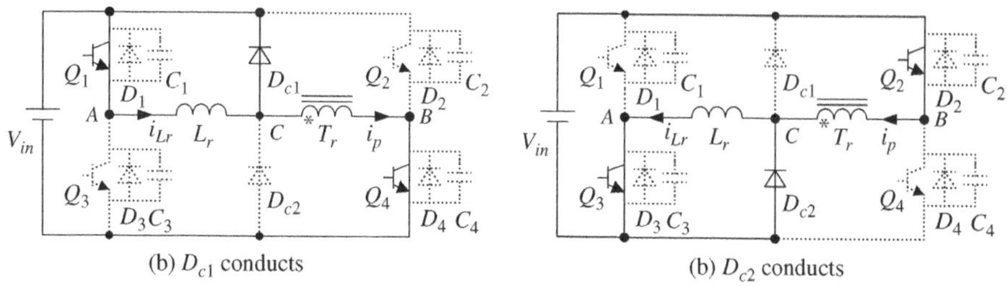

Figure 6.18 Equivalent circuits when the clamping diode conducts in the Tr-lag-type full-bridge converter

resonant inductor is shorted, and it is impossible for the transformer to be shorted by the clamping diode, as occurs in the Tr-lead-type converter. Therefore, as in the Tr-lead-type converter, if C_b is in series with the resonant inductor then the dc component across it will result in asymmetry of the resonant inductor current, but if C_b is in series with the transformer then the dc component across it will not result in asymmetry of the primary current, because the transformer will not be shorted by the clamping diodes.

The asymmetry of the primary and the resonant inductor current will degrade the reliability of the full-bridge converter. For the four converters, the best scheme is to insert a blocking capacitor in series with the transformer of the Tr-lag-type converter, as shown in Figure 6.15d.

6.7 Experimental Verification

In order to verify the operating principle of the ZVS PWM full-bridge converter with clamping diodes and compare the four kinds of topology shown in Figure 6.15, a 3 kW prototype was constructed with the following specifications:

- input voltage $V_{in} = 270\,\text{V} \pm 10\%$;
- output voltage $V_o = 28.5$ V; and
- output current $I_o = 100$ A.

The main parameters were:

- switches $Q_1(D_1 \; and \; C_1) \sim Q_4(D_4 \; and \; C_4)$: SPW47N60S5 (47 A/650 V);
- rectifier diodes D_{R1} and D_{R2}: DSEI2 × 121-02A;
- clamping diodes D_{c1} and D_{c2}: DSEI12-06A;
- resonant inductor $L_r = 4\,\mu\text{H}$;
- blocking capacitor $C_b = 5\,\mu\text{F}$;
- transformer turns ratio $K = 6.5$;

- output filter inductor $L_f = 3\,\mu\text{H}$;
- output filter capacitor $C_f = 2200\,\mu\text{F} \times 6$; and
- switching frequency $f_s = 100\,\text{kHz}$.

Figure 6.19 shows the experimental waveforms of i_p, i_{Lr}, i_{Dc1}, i_{Dc2}, v_{AB}, and v_{rect} (from the top to the bottom) of the four converters (see Figure 6.15) at full load under the nominal input voltage. It can be seen that almost no voltage oscillation occurs in v_{rect} thanks to the two clamping diodes. The clamping diode conducts twice in the Tr-lead-type converter and only once in the Tr-lag-type converter. Also, i_{Lr} in the zero state is smaller in the Tr-lag-type converter than it is in the Tr-lead-type converter.

The dc component of v_{AB} causes asymmetry of the resonant inductor current in positive and negative directions for both the Tr-lead-type and Tr-lag-type full-bridge

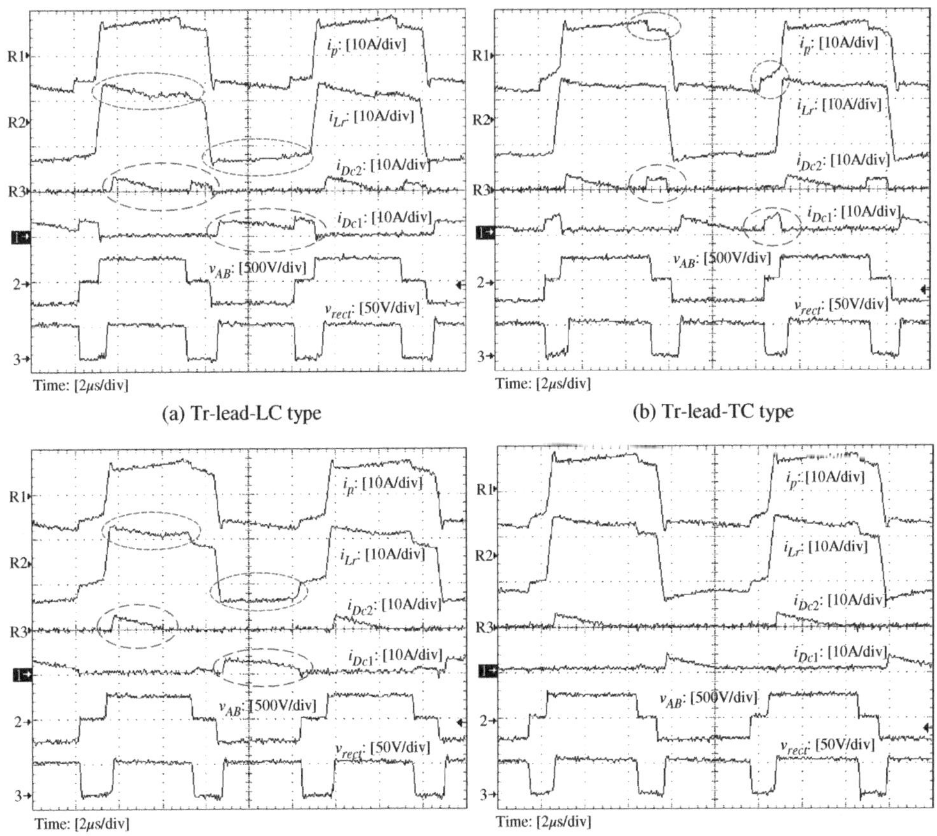

Figure 6.19 Experimental waveforms of four full-bridge converters

converters when the blocking capacitor is in series with the resonant inductor, as shown in Figure 6.19a,c. For the Tr-lead-type converter, the dc component of v_{AB} causes asymmetry of the transformer primary current, and as a result the currents of the two clamping diodes are asymmetrical, as shown in Figure 6.19b. For the Tr-lag-type converter, the dc component of v_{AB} will not result in asymmetry of the transformer primary current when the blocking capacitor is in series with the transformer. The transformer current is symmetrical, and the currents of the two clamping diodes are accordingly symmetrical too. Figure 6.19 clearly illustrates that the converter shown in Figure 6.15d is the preferred scheme.

Figure 6.20 shows the gate drive signal v_{GS}, the voltage across the drain and the source v_{DS}, and the drain current i_D of the leading-leg switch Q_1 and lagging-leg switch Q_2. This illustrates that all the switches turn off at zero voltage thanks to their junction capacitors. Before these switches are turned on, the voltages across the drain and source are clamped at zero by the conduction of their antiparallel diodes. Therefore, both the leading-leg and lagging-leg switches realize ZVS.

Figure 6.21 shows the efficiency of the Tr-lead-type and Tr-lag-type full-bridge converters. Figure 6.21a shows the efficiency versus the output current under a nominal input voltage of 270 V. Figure 6.21b shows the efficiency at full load of 100 A versus the input voltages. It can be seen that the efficiency decays with increasing input voltage under full-load conditions. The reason for this is that the higher the input voltage, the longer the duration of the zero state, resulting in a larger conduction loss at the primary side. As shown in Figure 6.21, the efficiency of the Tr-lag-type converter is higher than that of the Tr-lead-type converter, because the conduction loss is reduced due to the reduced resonant inductor current during the zero state. The efficiency of the Tr-lag-type full-bridge converter under full load is about 93.5%.

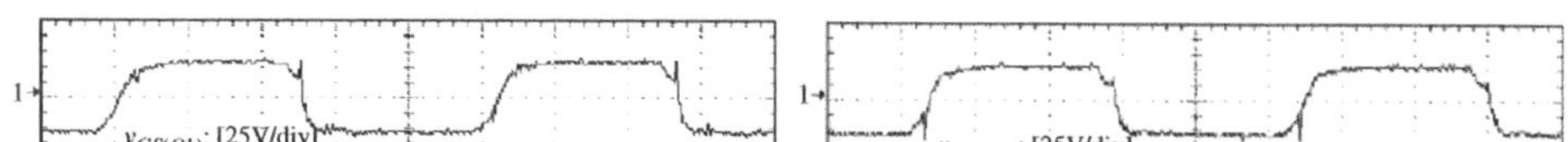

(a) Leading-leg switch Q_1 (b) Lagging-leg switch Q_2

Figure 6.20 v_{GS}, v_{DS}, and i_D waveforms of switches

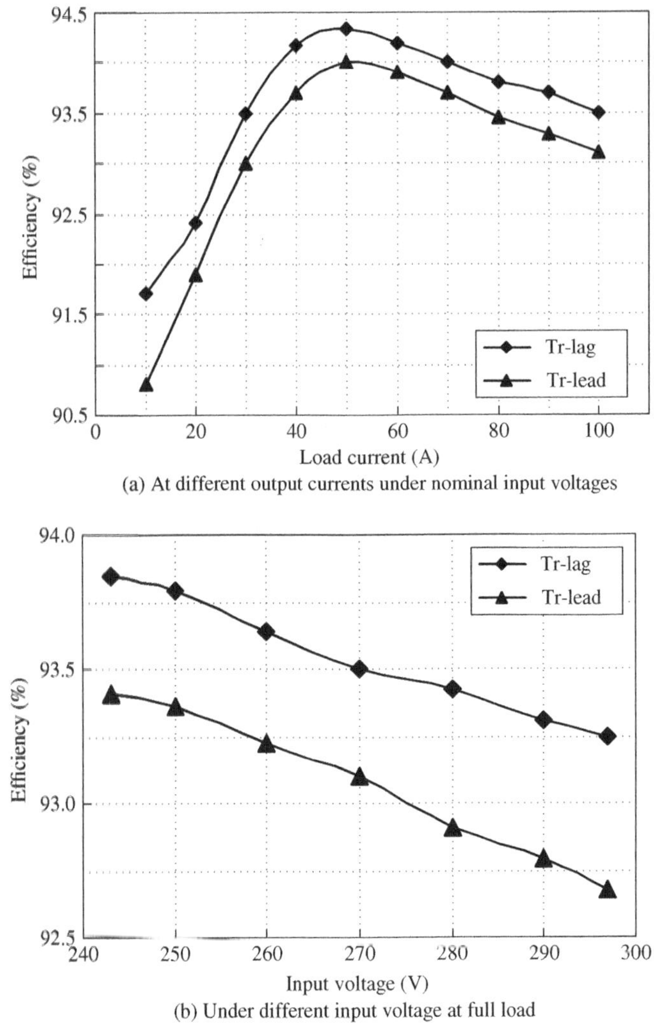

Figure 6.21 Measured overall conversion efficiencies of ZVS PWM full-bridge converters

6.8 Summary

In a ZVS PWM full-bridge converter, the resonance between the resonant inductor and the junction capacitor of the output rectifier diodes results in voltage oscillation and voltage spike across the output rectifier diode. Two clamping diodes are introduced to eliminate the voltage oscillation. The operation principle of the full-bridge converter with clamping diodes in which the transformer is connected to the leading leg is different from that for the converter in which the transformer connected to the lagging leg. The operating principle of the Tr-lead-type and Tr-lag-type full-bridge converters has been analyzed in detail in this chapter, and the following conclusions can be drawn:

1. The introduction of clamping diodes can effectively eliminate the voltage oscillation regardless of whether the transformer is connected to the leading or the lagging leg.
2. The clamping diode in the Tr-lead-type full-bridge converter conducts twice in a switching cycle, whereas that in the Tr-lag-type full-bridge converter conducts only once, which reduces the clamping diode current stress.
3. Compared with the Tr-lead-type full-bridge converter, the leading-leg switches in the Tr-lag-type full-bridge converter can more readily achieve ZVS, while the lagging-leg switches can less readily achieve it.
4. Compared with the Tr-lead-type full-bridge converter, the resonant inductor current of the Tr-lag-type full-bridge converter is smaller in the zero state, leading to a higher efficiency.
5. The duty cycle loss in the Tr-lag-type full-bridge converter is slightly smaller than that in the Tr-lead-type full-bridge converter.

Four kinds of topology with different blocking capacitor locations were compared and the Tr-lag-type full-bridge converter with the blocking capacitor connected in series with the transformer was found to be the preferred scheme.

References

1. Fisher, R.A. and Ngo, K.D.T. (1988) A 500 kHz, 250 W dc-dc converter with multiple outputs controlled by phase-shifted PWM and magnetic amplifiers. Proceedings of the High Frequency Power Conversion Conference, pp. 100–110.
2. Mweece, L.H., Wright, C.A. and Schlecht, M.F. (1989) A 1 kW 500 kHz front-end converter for a distributed power supply system. Proceedings of the IEEE Applied Power Electronics Conference and Exposition, pp. 423–432.
3. Redl, R., Sokal, N.O. and Balogh, L. (1991) A novel soft-switching full-bridge dc/dc converter: analysis, design considerations and experimental results at 1.5 kW, 100 kHz. *IEEE Trans. Power Electron.*, **26** (3), 408–418.
4. Redl, R., Balogh, L. and Edwards, D.W. (1993) Switch transitions in the soft-switching full-bridge PWM phase-shift dc/dc converter: analysis and improvements. Proceedings of the International Telecommunication Energy Conference, pp. 350–357.
5. Redl, R., Balogh, L. and Edwards, D.W. (1994) Optimum ZVS full-bridge dc/dc converter with PWM phase-shifted control: analysis, design considerations, and experimental results. Proceedings of the IEEE Applied Power Electronics Conference and Explosion, pp. 159–165.
6. Ruan, X. and Liu, F. (2004) An improved ZVS PWM full-bridge converter with clamping diodes. Proceedings of the IEEE Power Electronics Specialists Conference, pp. 1476–1481.
7. Sabate, J.A., Vlatkovic, V., Ridley, R.B. and Lee, F.C. (1991) High-voltage, high power, ZVS, full-bridge PWM converter employing an active snubber. Proceedings of the Virginia Power Electronics Center Seminar, pp. 125–130.
8. Kim, I.D., Nho, E.C. and Cho, G.H. (1992) Novel constant frequency PWM dc/dc converter with zero-voltage-switching for both primary switches and secondary rectifying diodes. *IEEE Trans. Ind. Electron.*, **39** (5), 444–452.

[illegible]

1. The introduction of clamping diodes [illegible] lation regardless of whether the [illegible]

[illegible]

7

Zero-Voltage-Switching PWM Full-Bridge Converters with Current Transformer to Reset the Clamping Diode Currents

7.1 Introduction

In zero-voltage-switching (ZVS) pulse-width-modulation (PWM) full-bridge converters, the resonant inductor (including the leakage inductor of the transformer) is used to realize ZVS for the power switches. When the full-bridge converter transits from zero state ($v_{AB}=0$) to +1 state ($v_{AB}=+V_{in}$) or −1 state ($v_{AB}=-V_{in}$), the output rectifier diodes begin to commutate and the resonant inductor resonates with the junction capacitor of the output rectifier diode, resulting in voltage oscillation and voltage spike across the output rectifier diodes. In Chapter 6, the voltage oscillation and voltage spike across the output rectifier diodes were effectively eliminated through the introduction of two clamping diodes at the primary side of the full-bridge converter, while the advantages of the ZVS PWM full-bridge converter were maintained.

From Figures 6.9 and 6.12, it can be seen that when the full-bridge converter transits from zero state to +1 or −1 state, the clamping diode conducts and the voltage oscillation across the output rectifier diode is eliminated. The current flowing through the clamping diode is equal to the difference between the resonant inductor current and the primary current of the transformer, and its initial value is equal to the peak resonant current with magnitude V_{in}/Z_r, where Z_r is the characteristic impedance of the resonant inductance and junction capacitor of the rectifier diode reflected to the primary side. The clamping diode current decays with the increase of the output filter inductor current. The output filter inductor is always designed to be larger, in order to achieve a lower current ripple, so the rate of increase of its current is very small, leading to a

Soft-Switching PWM Full-Bridge Converters: Topologies, Control, and Design, First Edition. Xinbo Ruan.

longer conduction time of the clamping diodes and relatively larger conduction losses in the clamping diodes, resonant inductor, and leading-leg switches. The conduction time of the clamping diodes will reach half of the switching period if the increase in the output filter inductor current is smaller than the peak resonant current in +1 or −1 state.

The analyses of the operation of the clamping diodes given in references [1–4] were all carried out under heavy load conditions. At light load, especially at no load, the operation of the two clamping diodes is very different. The two clamping diodes keep conducting alternately for almost half of the switching period, resulting in considerable conduction loss. At the same time, the clamping diodes may be hard turned off, with severe reverse recovery. Therefore, the clamping diodes can be easily damaged under light load conditions, especially in high-input-voltage applications [5].

It is necessary to make the clamping diode current decay rapidly to zero in order to reduce the conduction losses on the leading-leg switches, resonant inductor, and clamping diodes, and especially to avoid the hard turn-off of the clamping diodes under light load conditions.

This chapter first analyzes the operating principle of the ZVS PWM full-bridge converter with clamping diodes – especially the operation of the clamping diodes – under light load conditions. Several clamping diode current-reset schemes are proposed and compared. The operating principle of a full-bridge converter that employs a current transformer (CT) to reset the clamping diode current is analyzed. The clamping diode current-reset scheme with CT can make the clamping diode current decay rapidly over the full load range, while the excessive energy (the energy corresponding to the resonant current caused by the resonant inductor and junction capacitor of the output rectifier diode) stored in the resonant inductor is delivered to the input voltage source and the conversion efficiency is increased [6, 7]. Finally, an experimental prototype with 54 V/20 A output is used to demonstrate the operation of the clamping diode current-reset scheme with CT.

7.2 Operating Principle of the ZVS PWM Full-Bridge Converter with Clamping Diodes under Light Load Conditions

This section analyzes the operating principle of the ZVS PWM full-bridge converter with clamping diodes under light load conditions. Chapter 6 discussed the fact that in the full-bridge converter the transformer can be connected with the leading or the lagging leg and that the conversion efficiency of the full-bridge converter with the latter connection is relatively higher. Therefore, the Tr-Lag-type full-bridge converter is taken as an example for analysis. For convenience, the full-bridge converter is redrawn as shown in Figure 7.1. The key waveforms of the full-bridge converter with clamping diodes under light load conditions are shown in Figure 7.2 (there are two cases, depending on the load); the details of switching transitions are not shown explicitly because the intention is to describe the operating principle of the clamping diodes. To

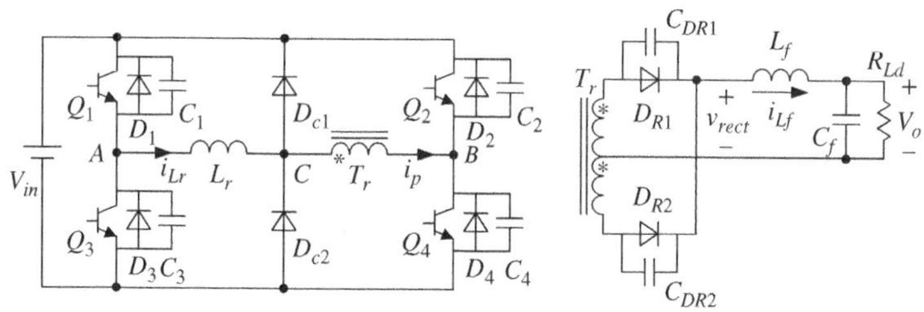

Figure 7.1 ZVS PWM full-bridge converter with clamping diodes

simplify the analysis, the following assumptions are made: all of the power switches, diodes, capacitors, inductors, and transformers are ideal, except for the output rectifier diodes, each of which is equivalent to an ideal diode and a parallel capacitor in order to simulate the reverse recovery; and $C_{DR1} = C_{DR2} = C_{DR}$.

The equivalent circuits of all switching modes are shown in Figure 7.3. Prior to t_0, Q_1 and Q_4 conduct, D_{R1} conducts, and D_{R2} is reversely biased, and the power is transferred from the input voltage source to the load, as shown in Figure 7.3a. The output rectified voltage v_{rect} equals V_{in}/K, where K is the primary-to-secondary-windings-turns ratio of the transformer. The primary current i_p equals the output filter inductor current i_{Lf} reflected to the primary side; that is, $i_p = i_{Lf}/K$. Thus, i_p increases linearly with the linear increase of i_{Lf}. The resonant inductor current i_{Lr} stays almost unchanged, and the difference between i_{Lr} and i_p flows through D_{c1} (the reason why D_{c1} conducts will be explained later).

At t_0, Q_1 is turned off and Q_3 is turned on with hard switching due to light load, as shown in Figure 7.3b. As D_{c1} and Q_4 still conduct, the transformer primary winding voltage $v_{CB} = V_{in}$, while i_{Lf} and i_p continue to increase linearly. V_{in} is negatively applied to L_r, forcing i_{Lr} to decline linearly. At t_1, i_{Lr} reduces to i_p and D_{c1} turns off naturally.

After t_1, L_r resonates with C_{DR2} and i_{Lr} decreases, as shown in Figure 7.3c. The further simplified equivalent circuit is shown in Figure 7.4a, where C'_{DR} and I'_{Lf} are the reflected C_{DR2} and $I_{Lf}(t_1)$ to the primary side, respectively. According to Figure 7.3c, we have:

$$i_{DR1}(t) + i_{CDR2}(t) = i_{Lf}(t) \tag{7.1}$$

$$i_{DR1}(t) - i_{CDR2}(t) = Ki_p(t) \tag{7.2}$$

According to Equations 7.1 and 7.2, i_{DR1} and i_{CDR2} can be derived as:

$$i_{DR1}(t) = \frac{1}{2}\left(i_{Lf}(t) + Ki_p(t)\right) \tag{7.3}$$

$$i_{CDR2}(t) = \frac{1}{2}\left(i_{Lf}(t) - Ki_p(t)\right) \tag{7.4}$$

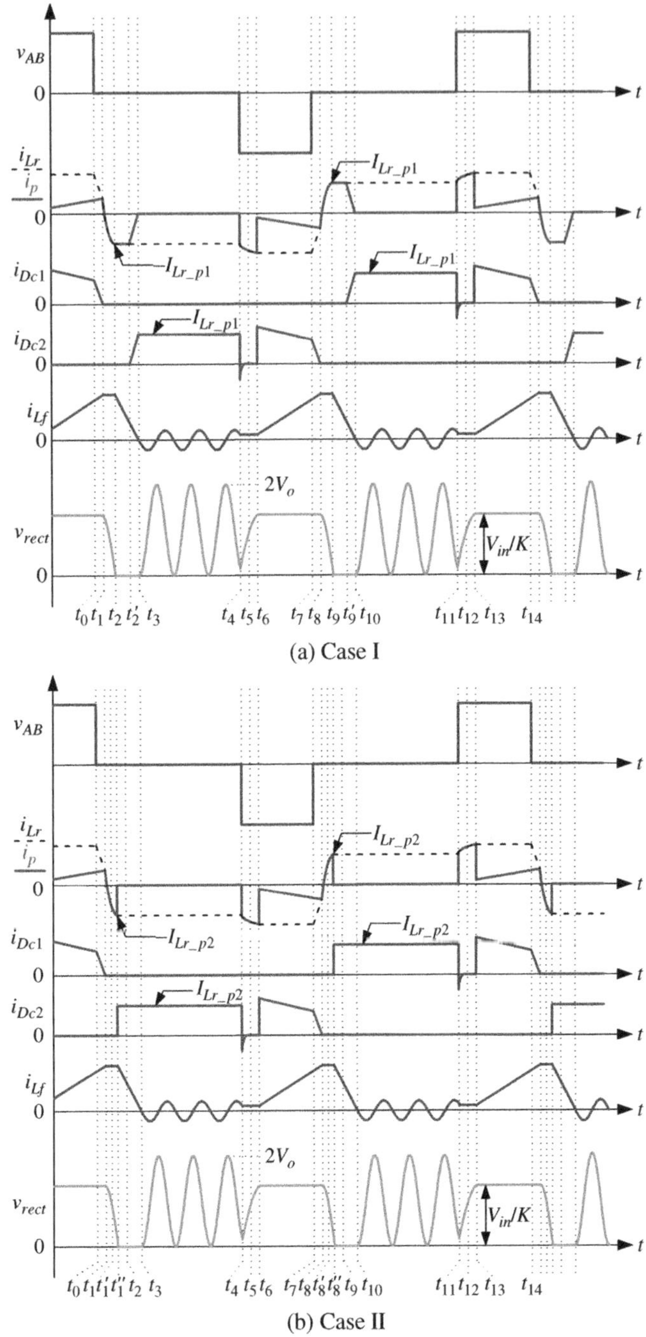

Figure 7.2 Key waveforms of the ZVS PWM full-bridge converter with clamping diodes under light load conditions

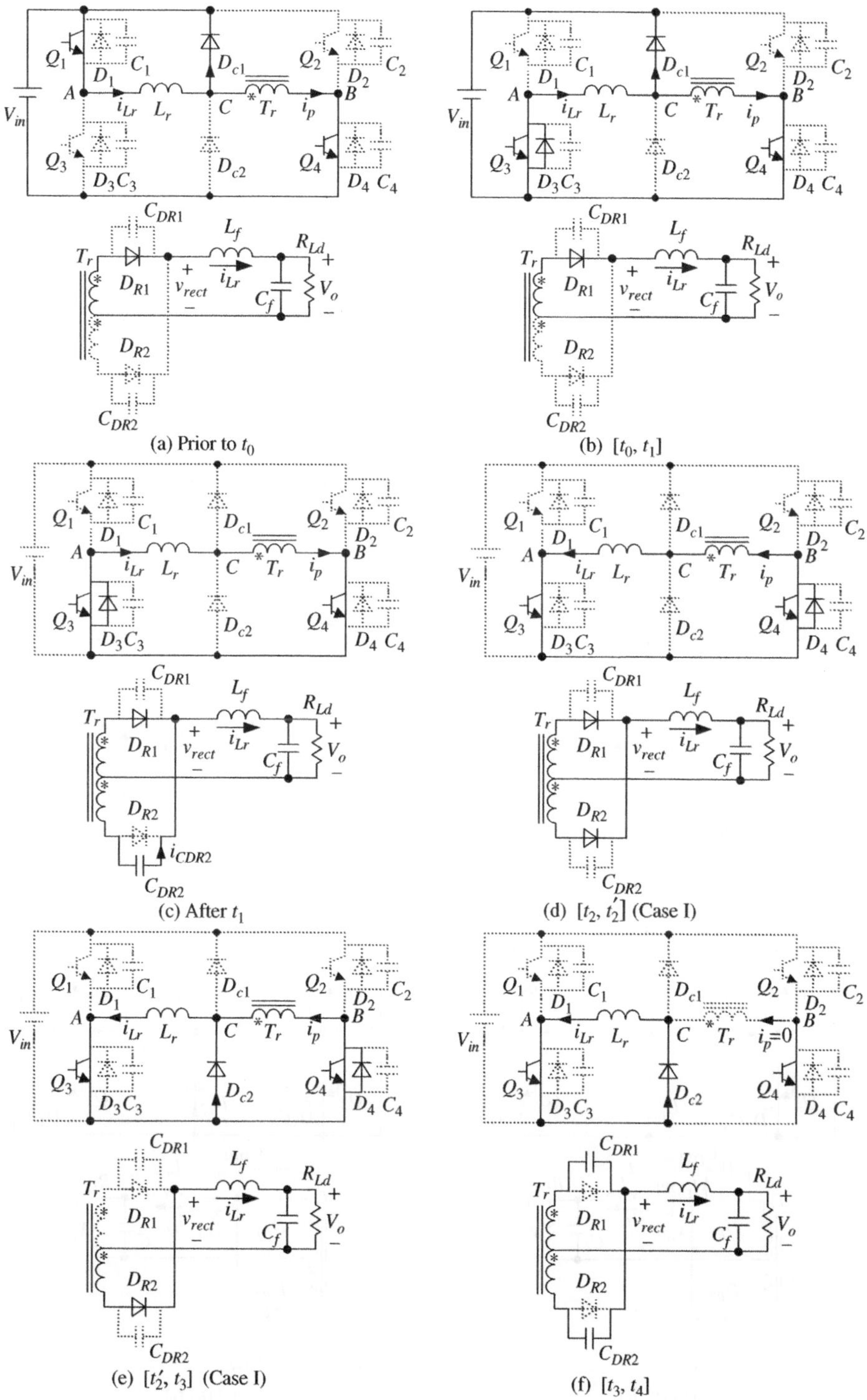

(a) Prior to t_0

(b) $[t_0, t_1]$

(c) After t_1

(d) $[t_2, t_2']$ (Case I)

(e) $[t_2', t_3]$ (Case I)

(f) $[t_3, t_4]$

Figure 7.3 Equivalent circuits of the full-bridge converter with clamping diodes under light load conditions

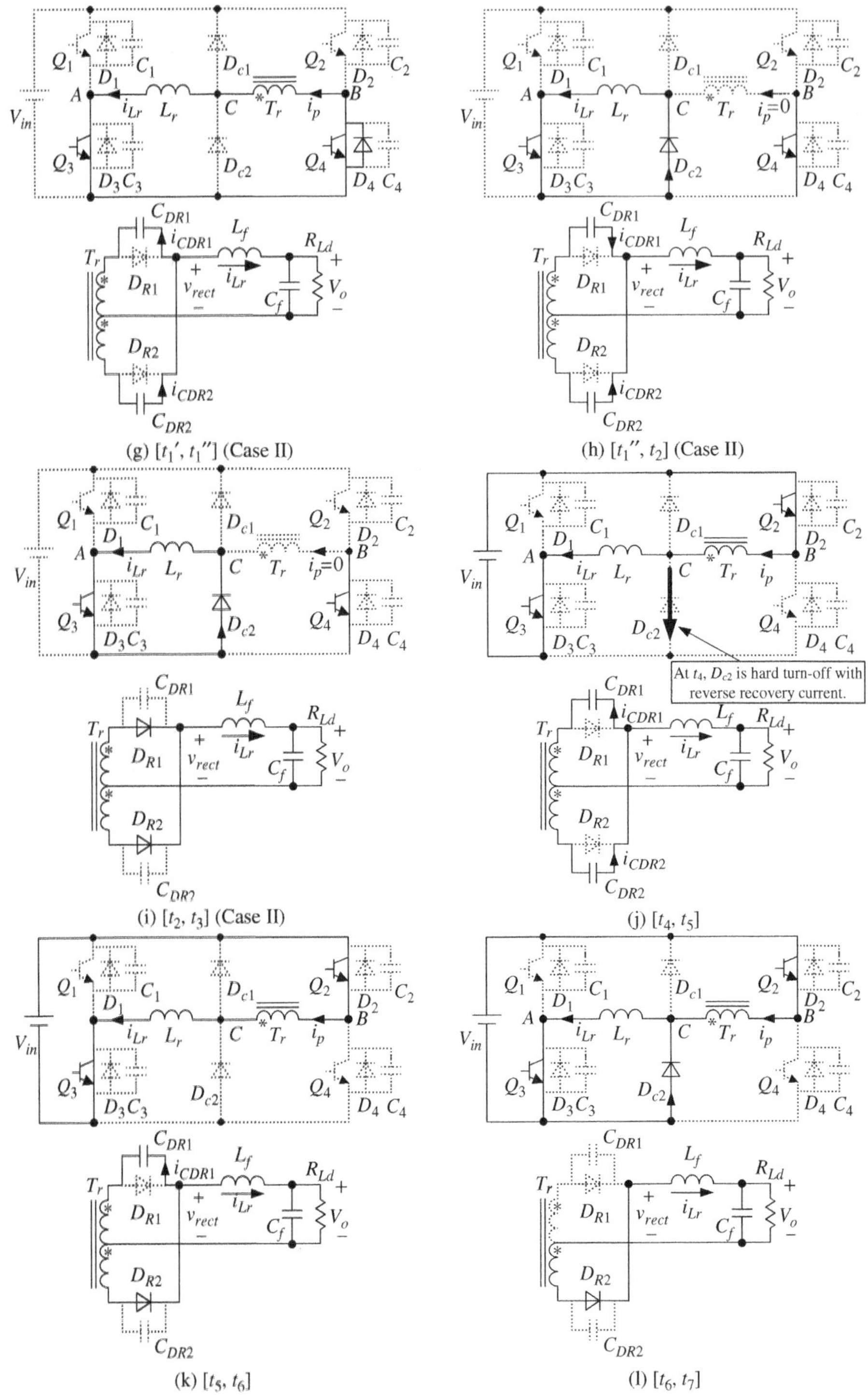

(g) $[t_1', t_1'']$ (Case II)

(h) $[t_1'', t_2]$ (Case II)

(i) $[t_2, t_3]$ (Case II)

(j) $[t_4, t_5]$

(k) $[t_5, t_6]$

(l) $[t_6, t_7]$

Figure 7.3 *(Continued)*

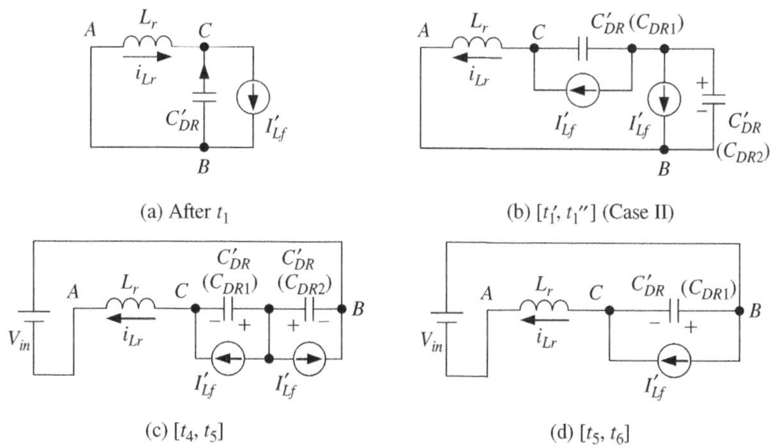

Figure 7.4 Further simplified equivalent circuits

According to Figure 7.4a, ignoring the slight change of i_{Lf}, i_p and the voltage across C_{DR2} can be expressed as:

$$i_p(t) = i_{Lr}(t) = \frac{I_{Lf}(t_1)}{K} - \frac{V_{in}}{Z_{r1}} \sin \omega_1 (t - t_1) \tag{7.5}$$

$$v_{CDR2}(t) = \frac{2V_{in}}{K} \cos \omega_1 (t - t_1) \tag{7.6}$$

where $C'_{DR} = 4C_{DR}/K^2$, $\omega_1 = 1/\sqrt{L_r C'_{DR}}$, and $Z_{r1} = \sqrt{L_r/C'_{DR}}$.

Substitution of Equation 7.5 into Equations 7.3 and 7.4 leads to:

$$i_{DR1}(t) = I_{Lf}(t_1) - \frac{KV_{in}}{2Z_{r1}} \sin \omega_1 (t - t_1) \tag{7.7}$$

$$i_{CDR2}(t) = \frac{KV_{in}}{2Z_{r1}} \sin \omega_1 (t - t_1) \tag{7.8}$$

According to Equations 7.5 and 7.6, it is known that i_{Lr} is still positive when v_{CDR2} reduces to zero if $I_{Lf}(t_1)/K \geq V_{in}/Z_{r1}$. After this, the converter will operate in freewheeling state and there will be no current flowing through the clamping diodes, just as in the heavy load conditions described in Chapter 6, which are not discussed here.

There are two possible cases, depending on the value of $I_{Lf}(t_1)$ when $I_{Lf}(t_1)/K < V_{in}/Z_{r1}$. This is different from the case in Chapter 6 and will be discussed in this section.

7.2.1 Case I: $0.5V_{in}/Z_{r1} \leq I_{Lf}(t_1)/K < V_{in}/Z_{r1}$ (Referring to Figure 7.2a)

According to Equations 7.5–7.7, if $0.5V_{in}/Z_{r1} \leq I_{Lf}(t_1)/K < V_{in}/Z_{r1}$ then when v_{CDR2} reduces to zero at t_2, i_{DR1} will still be positive and i_{Lr} will equal i_p, which is negative; that is:

$$I_{Lr}(t_2) = \frac{I_{Lf}(t_1)}{K} - \frac{V_{in}}{Z_{r1}} \triangleq -I_{Lr\_p1} \tag{7.9}$$

where $I_{Lr\_p1} = \frac{V_{in}}{Z_{r1}} - \frac{I_{Lf}(t_1)}{K}$.

After t_2, D_{R2} begins to conduct, as shown in Figure 7.3d. The transformer primary winding voltage is zero because both of the output rectifier diodes conduct, so that $v_{rect} = 0$ and i_{Lf} decreases linearly. The voltage across L_r is zero, so i_{Lr} and i_p remain unchanged at $-I_{Lr\_p1}$, freewheeling through Q_3 and Q_4 (D_4). The currents flowing through the two rectifier diodes are:

$$i_{DR1}(t) = \frac{1}{2}\left(i_{Lf}(t) + Ki_p(t)\right) = \frac{1}{2}\left(i_{Lf}(t) - KI_{Lr\_p1}\right) \tag{7.10}$$

$$i_{DR2}(t) = \frac{1}{2}\left(i_{Lf}(t) - Ki_p(t)\right) = \frac{1}{2}\left(i_{Lf}(t) + KI_{Lr\_p1}\right) \tag{7.11}$$

i_{DR1} and i_{DR2} decrease linearly with the linear decrease of i_{Lf}. At t_2', i_{Lf} decreases to KI_{Lr_p1}, i_{DR1} decreases to zero, and D_{R2} continues to conduct.

After t_2', both i_p and i_{Lf} continue decreasing and $i_p = -i_{Lf}/K$, the absolute value of i_p is smaller than that of i_{Lr}, and D_{c2} conducts, clamping the voltage of node C at zero, as shown in Figure 7.3e. Q_3 and D_{c2} conduct, i_{Lr} remains unchanged, and the difference between i_{Lr} and i_p flows through D_{c2}.

At t_3, i_{Lf} decreases to zero and D_{R2} turns off. Then i_{Lr} freewheels through Q_3 and D_{c2}, L_f resonates with C_{DR1} paralleling with C_{DR2}, and the transformer primary winding voltage and current are both zero, as shown in Figure 7.3f.

7.2.2 Case II: $I_{Lf}(t_1)/K < 0.5V_{in}/Z_{r1}$ (Referring to Figure 7.2b)

According to Equations 7.5–7.7, if $I_{Lf}(t_1)/K < 0.5V_{in}/Z_{r1}$ then when i_{DR1} decreases to zero at t_2, C_{DR2} is not completely discharged and i_p, i_{Lr}, and v_{CB} are expressed as:

$$I_{Lr}(t_1') = I_p(t_1') = -I_{Lf}(t_1)/K \tag{7.12}$$

$$V_{CB}(t_1') = \frac{KV_{CDR2}(t_1')}{2} = \sqrt{V_{in}^2 - \left(\frac{2I_{Lf}(t_1)\cdot Z_{r1}}{K}\right)^2} \tag{7.13}$$

Equation 7.12 implies that the direction of i_{Lr} and i_p is changed before t_1'. After t_1', D_{R1} turns off, C_{DR1} is charged, C_{DR2} is still discharged, and L_r resonates with C_{DR1} and C_{DR2}, leading to a decrease of v_{rect} and v_{CB}. i_p and i_{Lr} continue increasing in the negative direction, as shown in Figure 7.3g; the further simplified equivalent circuit is shown in Figure 7.4b, where C'_{DR} is the reflected C_{DR1} and C_{DR2} to the primary side and I_{Lf}' is $I_{Lf}(t_1)$ reflected to the primary side. According to Figure 7.4b, i_p, i_{Lr}, and v_{CB} are given by:

$$i_{Lr}(t) = i_p(t) = -\frac{V_{CB}(t_1')}{\sqrt{2}Z_{r1}}\sin\left(\sqrt{2}\omega_1\left(t-t_1'\right)\right) - \frac{I_{Lf}(t_1)}{K}\cos\left(\sqrt{2}\omega_1\left(t-t_1'\right)\right) \quad (7.14)$$

$$v_{CB}(t) = V_{CB}(t_1')\cos\left(\sqrt{2}\omega_1\left(t-t_1'\right)\right) - \frac{I_{Lf}(t_1)}{K}\sqrt{2}Z_{r1}\sin\left(\sqrt{2}\omega_1\left(t-t_1'\right)\right) \quad (7.15)$$

When v_{CDR2} reduces to equal v_{CDR1} at t_1'', both the primary and the secondary voltages of the transformer reduce to zero. Thus, the voltage of node C reduces to zero and D_{c2} conducts. At this point, i_{Lr} reaches its peak value of $-I_{Lr\_p2}$ and flows through Q_3 and D_{c2}. According to Equations 7.14 and 7.15, $-I_{Lr\_p2}$ can be expressed as:

$$-I_{Lr\_p2} = I_{Lr}(t_1'') = -\sqrt{\frac{V_{CB}^2(t_1')}{2Z_{r1}^2} + \frac{I_{Lf}^2(t_1)}{K^2}} \quad (7.16)$$

Substitution of Equation 7.13 into Equation 7.16 leads to:

$$I_{Lr\_p2} = \sqrt{\frac{V_{in}^2}{2Z_{r1}^2} - \frac{I_{Lf}^2(t_1)}{K^2}} \quad (7.17)$$

After t_1'', C_{DR1} and C_{DR2} share i_{Lf}, leading to a decrease in v_{rect}, with i_p remaining at zero. The equivalent circuit of the mode is shown in Figure 7.3h. At t_2, C_{DR1} and C_{DR2} are completely discharged, D_{R1} and D_{R2} share i_{Lf}, v_{rect} reduces to zero, and i_{Lf} decreases linearly, as shown in Figure 7.3i. At t_3, i_{Lf} decreases to zero. L_f then resonates with C_{DR1} paralleling with C_{DR2}, as shown in Figure 7.3f. It can be seen that no matter in case I or case II, the converter will operate in the switching mode with interval [t_3, t_4], as shown in Figure 7.3f.

According to Equation 7.17, we know that the smaller $I_{Lf}(t_1)$, the larger I_{Lr_p2}. Assuming that $I_{Lf}(t_1) = 0$, which means no-load conditions, we can get:

$$I_{Lr\_p2\max} = \frac{V_{in}}{\sqrt{2}Z_{r1}} \quad (7.18)$$

Based on the critical condition of case I and Equation 7.9, we get that the maximum value of I_{Lr_p1} is $0.5V_{in}/Z_{r1}$ and the minimum value is zero. Based on the critical condition of case II and Equation 7.17, the minimum and maximum values of $|i_{Lr}|$ are $0.5V_{in}/Z_{r1}$ and $V_{in}/(\sqrt{2}Z_{r1})$, respectively. From this analysis, we can conclude that under light load conditions the resonant inductor current will change its direction and freewheel during zero state ($v_{AB} = 0$), and that the smaller the load current, the larger the resonant inductor current.

In both case I and case II, prior to t_4, D_{c2} conducts and i_{Lr} remains almost unchanged at its maximum value. There is no current flowing through Q_4 (including its antiparallel diode D_4), although it is turned on. At t_4, Q_4 is turned off with ZVS/ZCS (zero-current-switching) and Q_2 is turned on with hard switching. v_{CB} cannot be changed suddenly, due to the existence of C_{DR1} and C_{DR2}, so the voltage of node C jumps to V_{in}, forcing the clamping diode current i_{Dc2} to decline acutely, as shown in Figure 7.3j. At this point, D_{c2} is hard turned off and suffers from serious reverse recovery current and massive losses, which can result in failure of the clamping diode in high-voltage applications. After D_{c2} is turned off, i_{Lr} equals i_p, C_{DR2} is discharged, and C_{DR1} is charged. The further equivalent circuit is shown in Figure 7.4c. It is worth noting that L_f resonates with C_{DR1} paralleling with C_{DR2} during $[t_3, t_4]$, and that at t_4, the voltages across C_{DR1} and C_{DR2} are the same, with the direction and magnitude of the voltages and of i_{Lf} depending on the time interval of $[t_3, t_4]$. It is assumed that i_{Lf} and the voltages across C_{DR1} and C_{DR2} are all positive in Figure 7.4c.

At t_5, C_{DR2} is completely discharged and D_{R2} conducts, and C_{DR1} is charged by i_{Lr}, as shown in Figure 7.3k. Its further equivalent circuit is shown in Figure 7.4d.

At t_6, the voltage across C_{DR1} is charged to $2V_{in}/K$, the voltage of node C reduces to zero, and D_{c2} conducts again, as shown in Figure 7.3l. i_p declines to the reflected filter inductor current, i_{Lr} remains almost unchanged, and their difference flows through D_{c2} (this is the reason the clamping diode conducts during the active state ($v_{AB} = +V_{in}$ or $-V_{in}$) under light load conditions). It can be seen from Figure 7.2 that the longest conduction time for the clamping diode is nearly half of the switching period.

From this analysis, it can be seen that the operating principle of the full-bridge converter with clamping diodes under light load conditions is remarkably different from that under heavy load conditions, and that the critical current is $I_{Lf}(t_1) = KV_{in}/Z_{r1}$. The primary current will change its direction during zero state ($v_{AB} = 0$) and the clamping diode will conduct for almost half of the switching period under light load conditions. Besides the unavoidable serious reverse recovery of the clamping diodes, the phenomenon also causes large conduction loss on the clamping diodes and leading-leg switches when the full-bridge converter operates under near-no-load conditions. It is therefore necessary to reset the clamping diode current to zero rapidly over the full load range.

7.3 Clamping Diode Current-Reset Scheme

7.3.1 Reset Voltage Source

It is known from Chapter 6 that D_{c1} conducts after the full-bridge converter transits from zero to +1 state, as shown in Figure 7.5a, and that D_{c2} conducts after it transits from zero to −1 state, as shown in Figure 7.5b (only the primary side circuits are shown, for simplicity). The current flowing through D_{c1} and D_{c2} is the difference between the resonant inductor current and the output filter inductor current reflected to

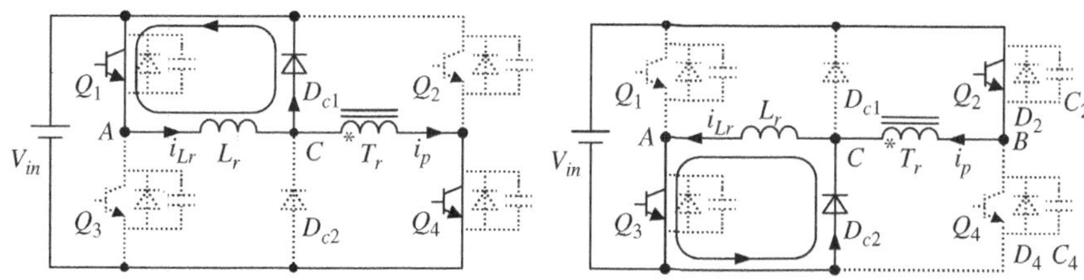

(a) +1 state (b) −1 state

Figure 7.5 Clamping diode conduction conditions

the primary side, and the initial clamping diode current is the resonant current, which is caused by the resonance between the resonant inductor and the junction capacitor of the output rectifier diode. The resonant inductor is shorted by the conduction of the clamping diode, and its current remains almost unchanged. The clamping diode current decreases linearly with the linear increase of the output filter inductor current, and the rate of decline of the clamping diode current depends on the rate of rise of the output filter inductor current. The output filter inductor is intentionally reduced to increase the output filter inductor current ripple in order to make the clamping diode current decrease rapidly. However, a larger output filter capacitor is required and the power density of the full-bridge converter is reduced.

In order to reset the clamping diode current rapidly, a reset voltage source can be brought into the current loop and connected in series with the resonant inductor or the clamping diodes, as shown in Figure 7.6.

Figure 7.6a shows that the reset voltage source v_{reset} is connected in series with the resonant inductor. The polarity of v_{reset} must comply with the direction of the resonant inductor current; that is, when the resonant inductor current i_{Lr} is positive, v_{reset} should be positive (the same direction as the reference voltage), and when i_{Lr} is negative, v_{reset} should be negative (the opposite direction to the reference voltage). When a clamping diode current loop is formed, v_{reset} is directly applied to L_r, forcing i_{Lr} to decrease rapidly and thus the clamping diode current to decline to zero rapidly.

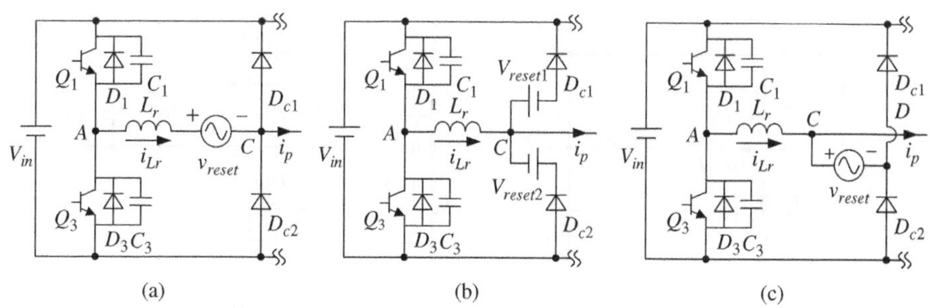

(a) (b) (c)

Figure 7.6 Three clamping diode current-reset schemes

Figure 7.6b shows that two independent reset voltage sources are connected in series with the clamping diodes. Each is unipolar because the clamping diode current is unidirectional. In practice, the two reset voltage sources can be implemented by a resistor and a Zener diode, respectively. The voltage drop on the resistor plays the role of the reset voltage source when the clamping diode current flows through the resistor. However, the excessive energy stored in the resonant inductor is dissipated in the resistor, leading to a reduced efficiency.

The two independent reset voltage sources shown in Figure 7.6b can be replaced by a single reset voltage source as they do not work at the same time, as shown in Figure 7.6c. As in Figure 7.6a, the polarity of the reset voltage source shown in Figure 7.6c must comply with the direction of the current flowing through it. Two concrete cases for Figure 7.6c are presented in reference [2], where either a resistor or two back-to-back Zener diodes are adopted as the alternative reset voltage source. However, all of the excessive energy stored in the resonant inductor will be dissipated in the resistor or Zener diodes.

7.3.2 Implementation of the Reset Voltage Source

For the reset voltage source shown in Figure 7.6a, an auxiliary winding can be added in the primary side of the transformer, as shown in Figure 7.7 [7]. When the full-bridge converter operates in +1 state, the voltages across the primary winding of the transformer and the auxiliary winding are $+V_{in}$ and $+V_{in} \cdot n_3/n_1$, respectively, where n_1 and n_3 are the turns of the primary and auxiliary winding, respectively. When the full-bridge converter operates in −1 state, the voltages across the primary winding of the transformer and the auxiliary winding are $-V_{in}$ and $-V_{in} \cdot n_3/n_1$, respectively. Hence, the voltage polarity of the auxiliary winding complies with the direction of the resonant inductor current. Therefore, the clamping diode current can be reset rapidly thanks to the rapid decrease in the resonant inductor current. However, the reset winding voltage is directly linked to the primary duty cycle and will lose its function when the primary duty cycle is relatively small. On the other hand, the resonant inductor current changes its direction and freewheels during the zero state ($v_{AB}=0$) under light

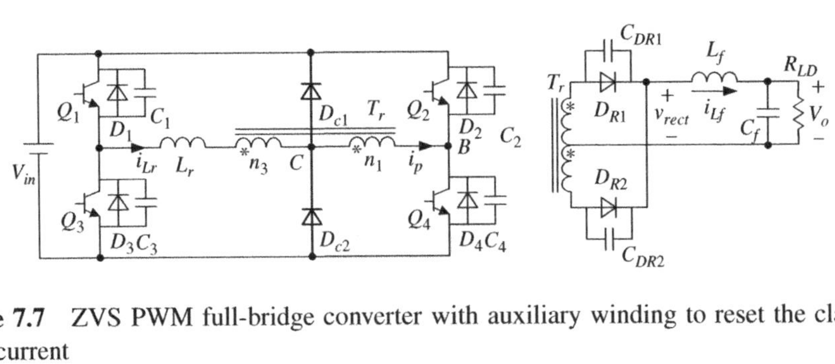

Figure 7.7 ZVS PWM full-bridge converter with auxiliary winding to reset the clamping diode current

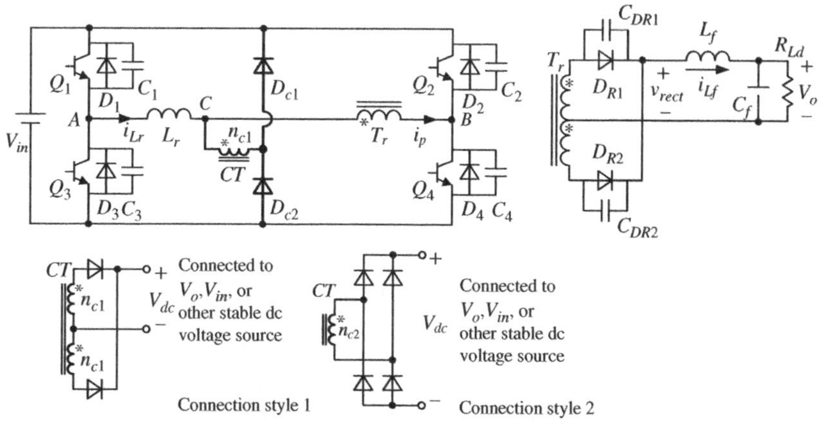

Figure 7.8 ZVS PWM full-bridge converter with CT to reset the clamping diode current

load conditions and the primary winding voltage of the transformer becomes zero. Hence, the reset winding cannot reset the clamping diode current during the zero state.

The reset voltage source in Figure 7.6c can be realized by a CT and related circuit, as shown in Figure 7.8 [8]. The CT primary side is connected to node C and the midpoint of the two clamping diodes. The secondary side of the CT is rectified and connected to a stable dc voltage source, which may be the output voltage V_o, the input voltage V_{in}, or another dc voltage source. Both the center-tapped rectifier and the full-bridge rectifier can be used for the secondary side of the CT; the former is suitable for a lower dc voltage and the latter for a higher one. When the resonant inductor current is positive, the CT primary side current flows into the dot side and through the clamping diode D_{c1}, while the CT secondary side current flows out from the dot side and into the connected dc source V_{dc}. The polarity of the dot side of the CT primary side is positive and the primary voltage is $V_{dc} \cdot n_{c1}/n_{c2}$, where n_{c1} and n_{c2} are turns of the CT primary and secondary windings, respectively. The CT primary side voltage makes the resonant inductor current decline, and the current flowing through D_{c1} is reset. Similarly, when the resonant inductor current is negative, the CT primary side current flows through D_{c2} and out from the dot side, while the CT secondary side current flows into the dot side and through the connected dc source V_{dc}. The polarity of the dot side of the CT primary side is then $-V_{dc} \cdot n_{c1}/n_{c2}$. The CT primary side voltage makes the resonant inductor current decline and the current flowing through D_{c2} is reset. It can be seen that the voltage polarity of the CT primary winding changes with the direction of the resonant inductor current, and a reset voltage source is obtained with the help of the CT secondary winding and its rectifier circuit.

When the output rectifier circuit of the CT secondary side is connected to the output voltage, the excessive energy stored in the resonant inductor can be directly delivered to load, resulting in improved conversion efficiency. However, during the starting stage and the short-circuit current-limiting state, the output voltage is lower than the

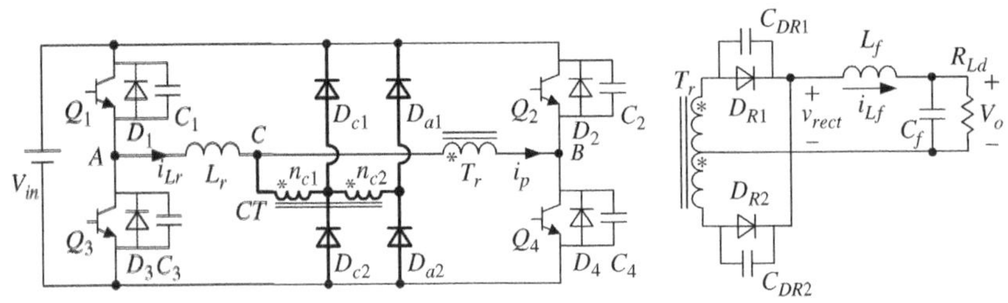

Figure 7.9 Simplified ZVS PWM full-bridge converter with CT

normal value, and the reflected voltage to the CT primary side is insufficient to reset the clamping diode current. This issue can be avoided by connecting the output rectifier circuit of the CT secondary side to the input voltage source. The clamping diode current reset time is constant even when the input voltage is changed, which will be explained later. Generally, the input voltage of the full-bridge converter is relatively higher; hence, a full-bridge rectifier circuit is employed for the CT secondary side; that is, connection style 2 in Figure 7.8. It can be seen from Figure 7.8 that when D_{c1} or D_{c2} conducts, the top or bottom rectifier diode of the left side also conducts, in the secondary connection style. Hence, the two left-side rectifier diodes can be replaced by two clamping diodes as galvanic isolation is not mandatory between the CT primary and secondary sides, so two rectifier diodes can be saved. The final circuit is shown in Figure 7.9, where D_{a1} and D_{a2} are the CT secondary rectifier diodes.

7.4 Operating Principle of the ZVS PWM Full-Bridge Converter with Current Transformer

The operating principle of the ZVS PWM full-bridge converter with CT as the clamping diode current-reset circuit will be discussed in this section. The converter circuit is shown in Figure 7.9. Converter operations under heavy and light load conditions are different and will be analyzed separately.

7.4.1 Operating Principle under Heavy Load Conditions

Figure 7.10 gives the key waveforms of the ZVS PWM full-bridge converter with CT operating under heavy load conditions. To simplify the analysis, the following assumptions are made:

1. All the switches and diodes are ideal, except for the output rectifier diode, which is equivalent to an ideal diode and a paralleled capacitor, simulating the reverse recovery; $C_{DR1} = C_{DR2} = C_{DR}$.

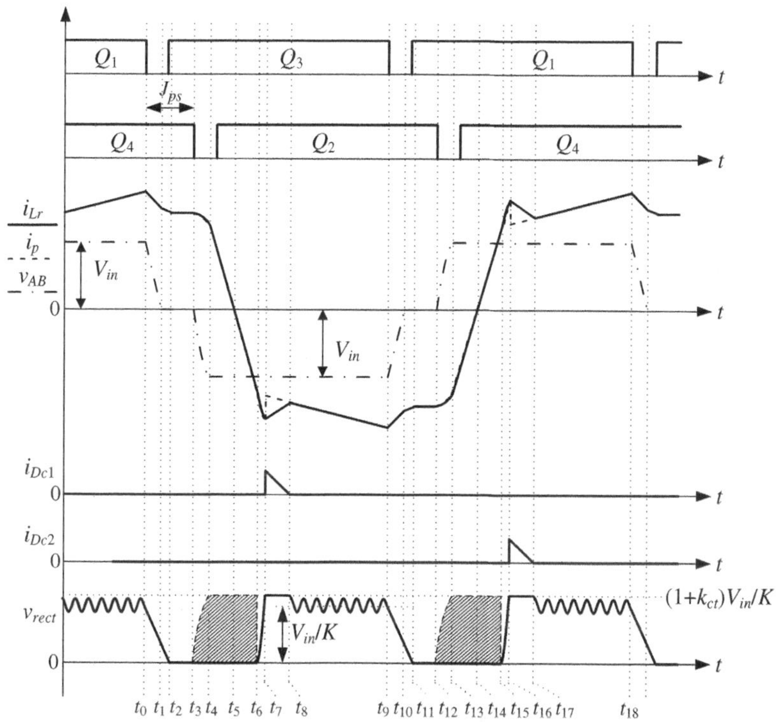

Figure 7.10 Key waveforms of the full-bridge converter with CT under heavy load conditions

2. All the capacitors, inductors, and transformer are ideal.
3. The turns ratio of the CT is $k_{ct} = n_{c1}/n_{c2}$.

There are 16 switching modes in a switching period. Since the first half-period is similar to the second, only the first is given here for brevity. The equivalent circuits of each mode are shown in Figure 7.11.

1. **Mode 1, $[t_0, t_1]$ (Figure 7.11a):** Prior to t_0, Q_1, Q_4, and D_{R1} conduct and D_{R2} is reversely blocked. The power is transferred from the input voltage source V_{in} to the load. At t_0, Q_1 is turned off at zero voltage thanks to C_1 and C_3, which limit the rate of rise of the voltage across Q_1. i_{Lr} charges C_1 and discharges C_3, and the potential voltage of point A decays. Meanwhile, the capacitor C_{DR2} is discharged. As the potential voltage of point C is greater than zero and lower than V_{in}, the two clamping diodes are both reverse-biased. The voltage of C_3 decreases to zero at t_1 and D_3 conducts naturally.
2. **Mode 2, $[t_1, t_2]$ (Figure 7.11b):** In this mode, Q_3 can be turned on at zero voltage. C_{DR2} continues to be discharged, since the voltage of point C is still higher than

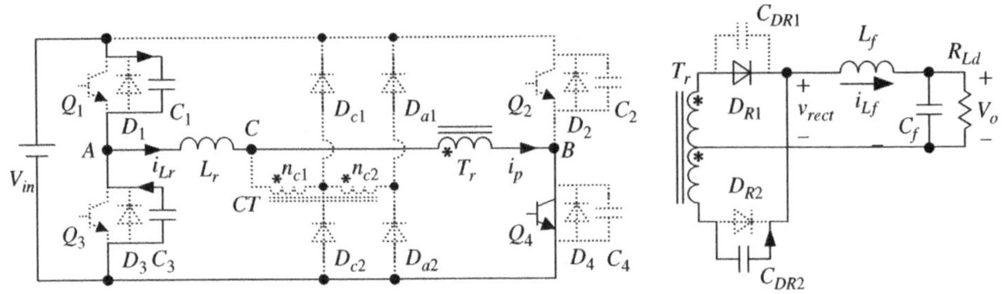

(a) $[t_0, t_1]$

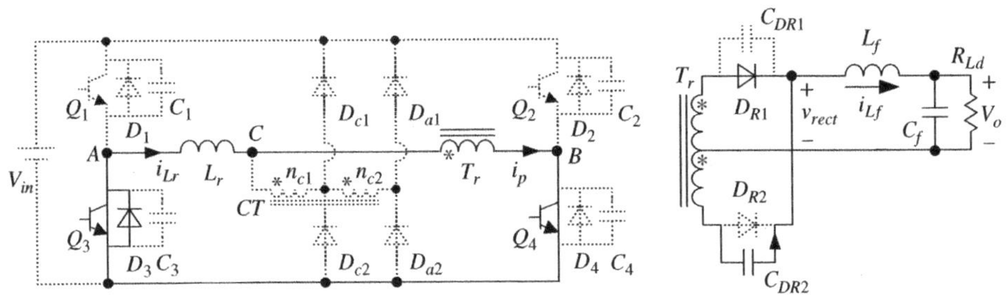

(b) $[t_1, t_2]$

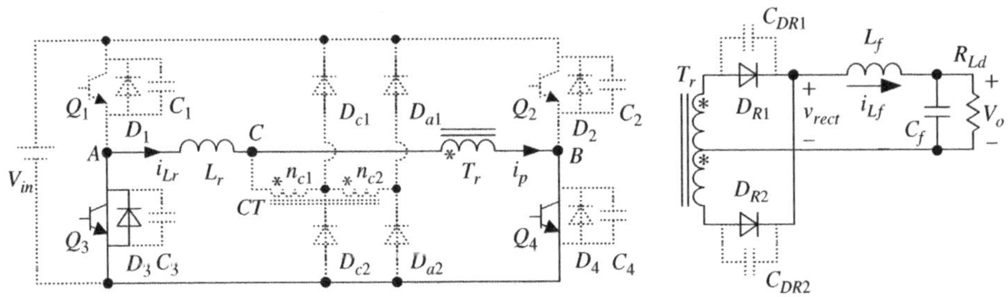

(c) $[t_2, t_3]$

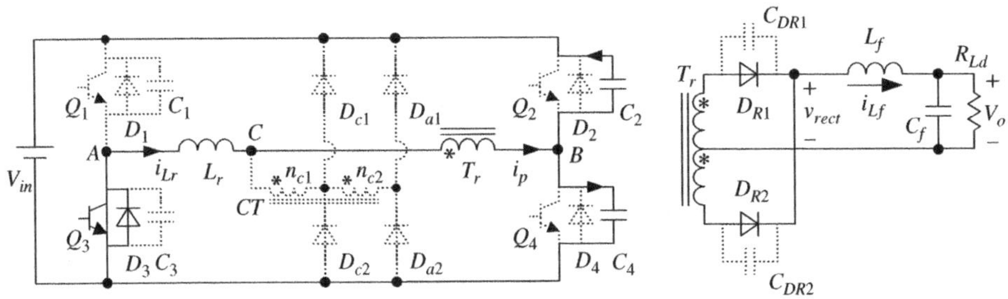

(d) $[t_3, t_4]$

Figure 7.11 Equivalent circuits of the full-bridge converter with CT under heavy load conditions

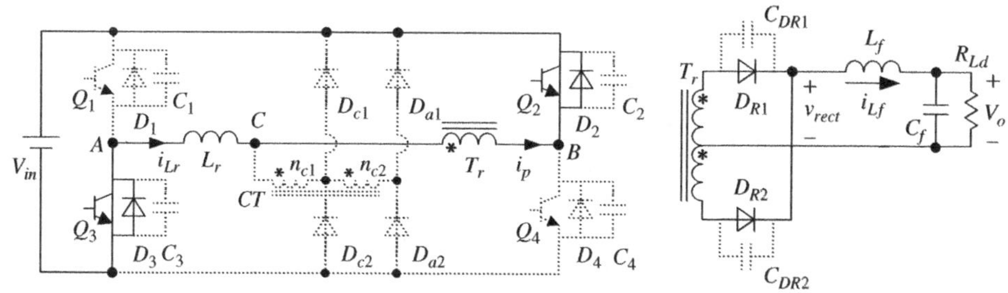

(a) $[t_4, t_6]$

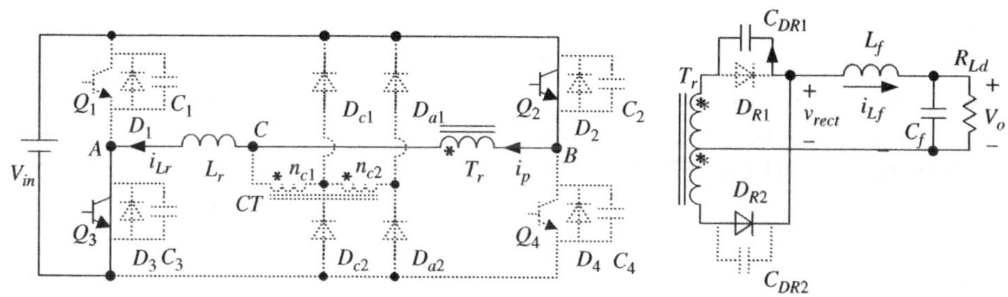

(b) $[t_6, t_7]$

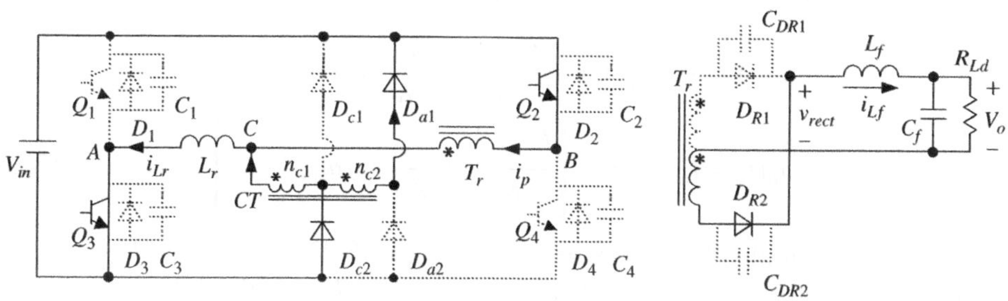

(c) $[t_7, t_8]$

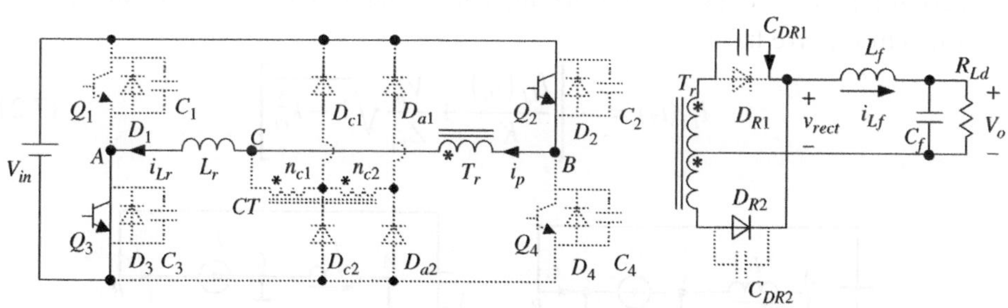

(d) $[t_8, t_9]$

Figure 7.11 (*Continued*)

zero. i_{Lr} and i_p continue decaying. At t_2, the voltage across C_{DR2} reduces to zero and D_{R2} begins to conduct, and the voltage of point C reduces to zero.

3. **Mode 3, [t_2, t_3] (Figure 7.11c):** In this mode, both D_{R1} and D_{R2} conduct, clamping the voltages of the transformer primary and secondary sides at zero. i_{Lr} is equal to i_p and the circuit operates in freewheeling state.
4. **Mode 4, [t_3, t_4] (Figure 7.11d):** At t_3, Q_4 is turned off at zero voltage thanks to C_2 and C_4, which limit the rate of rise of the voltage across Q_4. C_4 is charged and C_2 discharged in a resonant manner. At t_4, v_{C4} rises to V_{in} and v_{C2} reduces to zero, and D_2 conducts naturally.
5. **Mode 5, [t_4, t_6] (Figure 7.11e):** After D_2 conducts, Q_2 can be turned on at zero voltage. i_{Lr} is equal to i_p, and both decay linearly at the rate of V_{in}/L_r. At t_5, i_{Lr} and i_p cross zero and continue increasing linearly in the negative direction. The load current flows through both the output rectifier diodes. At t_6, i_p increases to the reflected current of i_{Lf}, so D_{R1} is turned off.
6. **Mode 6, [t_6, t_7] (Figure 7.11f):** After t_6, L_r resonates with C_{DR1}, which is charged in a resonant manner. i_p and i_{Lr} continue increasing and v_{rect} increases. The voltage across C_{DR1} increases and the voltage of node C is reduced accordingly, because the voltage of node B is V_{in}. The further simplified equivalent circuit of this mode is shown in Figure 7.12a, where C'_{DR} and I'_{Lf} are the reflected C_{DR1} and $I_{Lf}(t_6)$ to the primary side, respectively. v_{rect} and i_p are given by:

$$v_{rect}(t) = \frac{1}{2}v_{CDR1}(t) = \frac{v_{BC}(t)}{K} = \frac{V_{in}}{K}[1 - \cos\omega_1(t - t_6)] \tag{7.19}$$

$$i_p(t) = i_{Lr}(t) = -\left[\frac{I_{Lf}(t_6)}{K} + \frac{V_{in}}{Z_{r1}}\sin\omega_1(t - t_6)\right] \tag{7.20}$$

At t_7, the potential voltage of node C reduces to $-k_{ct}V_{in}$, so D_{c2} and D_{a1} conduct, the voltage across C_{DR1} is clamped at $2(1 + k_{ct})V_{in}/K$, and the voltage oscillation and spike across D_{R1} are eliminated. Meanwhile, v_{rect} is clamped at $(1 + k_{ct})V_{in}/K$ accordingly. i_{Lf} is considered to be unchanged due to the very short time interval of this mode. We have:

$$I_{Lr}(t_7) = -\left[\frac{I_{Lf}(t_6)}{K} + \frac{V_{in}}{Z_{r1}}\sqrt{1 - k_{ct}^2}\right] \tag{7.21}$$

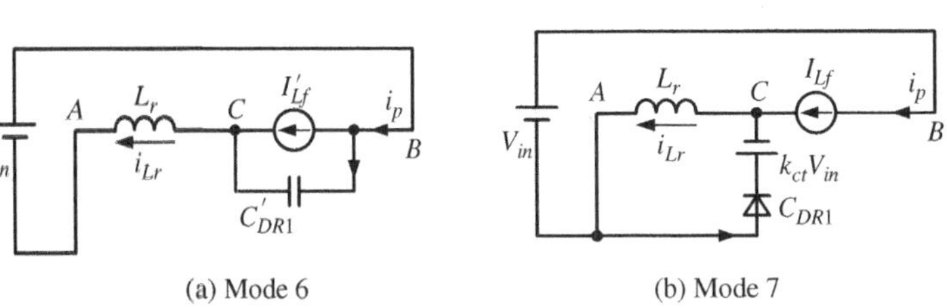

(a) Mode 6 (b) Mode 7

Figure 7.12 Further equivalent circuits of modes 6 and 7

7. **Mode 7, [t_7, t_8] (Figure 7.11g):** At t_7, as D_{c2} conducts, i_p declines downwards to the reflected output filter inductor current and increases in the negative direction. At the same time, i_{Lr} flows through Q_3, D_{c2}, and the primary winding of the CT. The CT secondary current flows through D_{c2} and D_{a1} into the input voltage source. V_{in} is applied on the CT secondary winding and the induced voltage across the CT primary winding is $-k_{ct}V_{in}$, which is applied to L_r, forcing i_{Lr} to decrease quickly. The further simplified equivalent circuit of this stage is shown in Figure 7.12b. At t_8, i_{Lr} decays to i_p, and D_{c2} and D_{a1} turn off naturally. In this mode, the current flows through D_{c2} and D_{a1} are:

$$i_{Dc2}(t) = (1 + k_{ct})\left(-I_{Lr}(t_7) - \frac{i_{Lf}(t)}{K}\right) \tag{7.22}$$

$$i_{Da1}(t) = k_{ct}\left(-I_{Lr}(t_7) - \frac{i_{Lf}(t)}{K}\right) \tag{7.23}$$

According to Equations 7.21 and 7.22, the peak value of i_{Dc2} is:

$$I_{Dc2}(t_7) = (1 + k_{ct})\frac{V_{in}}{Z_{r1}}\sqrt{1 - k_{ct}^2} \tag{7.24}$$

8. **Mode 8, [t_8, t_9] (Figure 7.11h):** At t_8, L_r resonates with C_{DR1}. The rectified voltage v_{rect} is given by:

$$v_{rect}(t) = \frac{V_{in}}{K}[1 + k_{ct}\cos\omega_1(t - t_8)] \tag{7.25}$$

Equation 7.25 shows that the maximum value of v_{rect} will never exceed $(1 + k_{ct})$ V_{in}/K, although a slight oscillation exists. In practice, v_{rect} will finally converge to the average value V_{in}/K, since the inherent parasitic resistor exists in the power stage, as will be shown in Section 7.6.

7.4.2 Operating Principle under Light Load Conditions

This section discusses the operating principle of the ZVS PWM full-bridge converter with CT under light load conditions. As in Section 7.2, the switching transitions are ignored in order to highlight the operation of the clamping diodes. The key waveforms of this converter are shown in Figure 7.13.

Prior to t_0, Q_1, Q_4, and D_{R1} conduct and D_{R2} is reversely blocked. The power is transferred from the input voltage source V_{in} to the load, as shown in Figure 7.14a. The difference from the full-bridge converter without CT is that the clamping diode does not conduct. At t_0, Q_1 is turned off and Q_3 is turned on, $v_{AB} = 0$, C_{DR2} resonates with the resonant inductor, C_{DR2} begins to be discharged, and the resonant inductor current decays, as shown in Figure 7.14b. This switching mode is the same as that in Figure 7.3c. As in Section 7.2, there are two cases, depending on the load; these will be discussed separately.

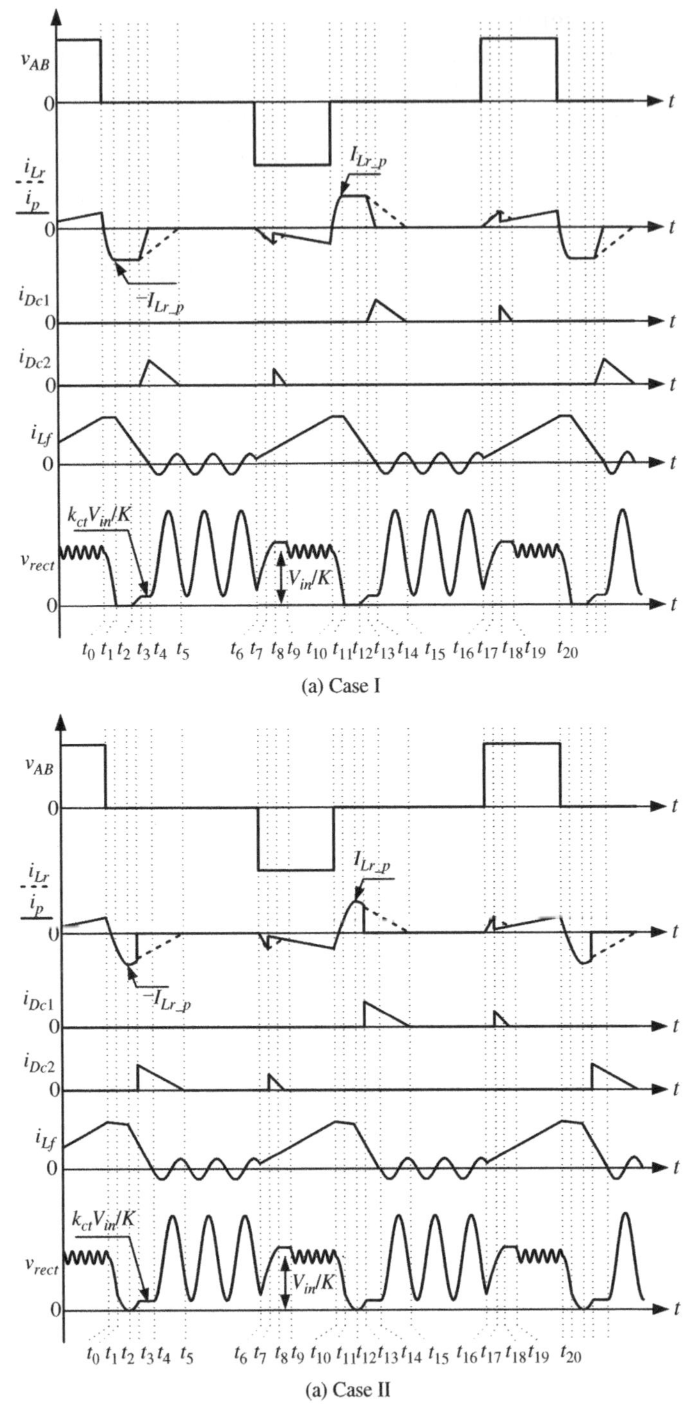

Figure 7.13 Key waveforms of the full-bridge converter with CT under light load conditions

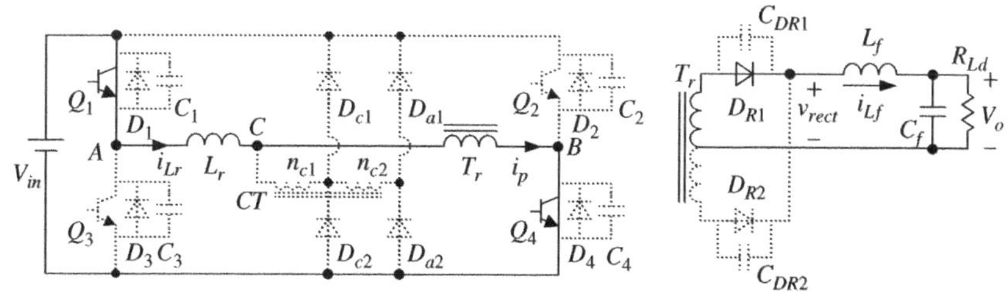

(a) Prior t_0

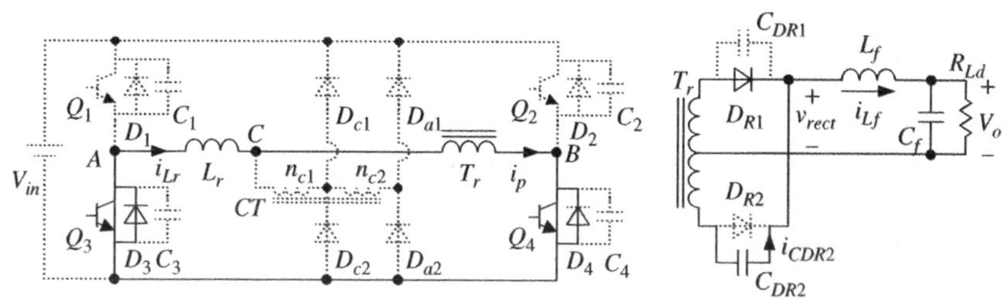

(b) After t_0

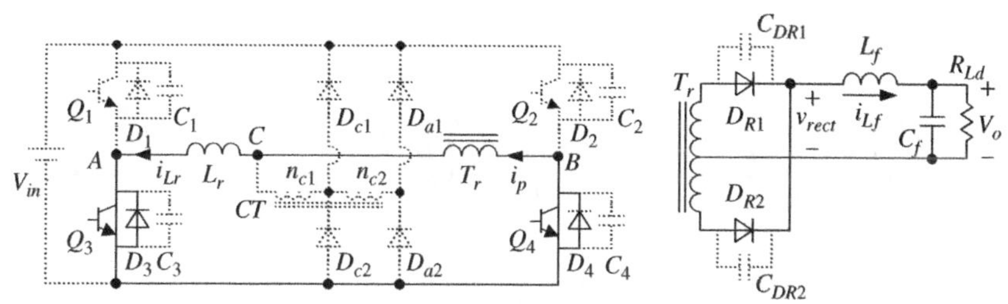

(c) $[t_1, t_2]$ (Case I)

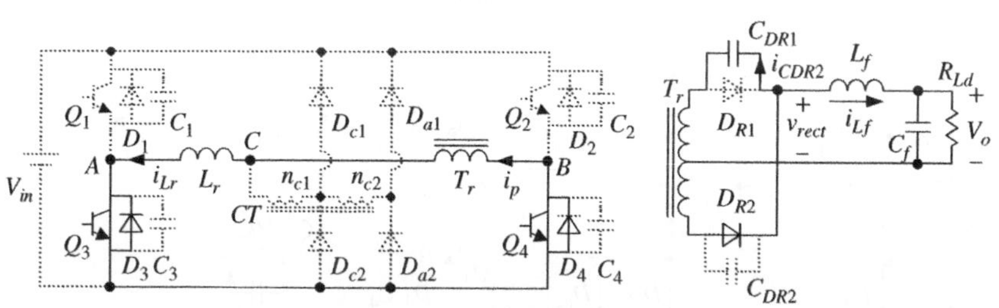

(d) $[t_2, t_3]$ (Case I)

Figure 7.14 Equivalent circuits of the full-bridge converter with CT under light load conditions

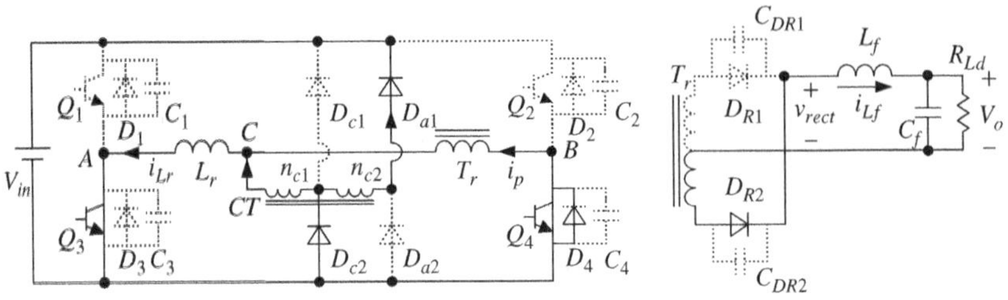

(e) $[t_3,t_4]$ (Case I)

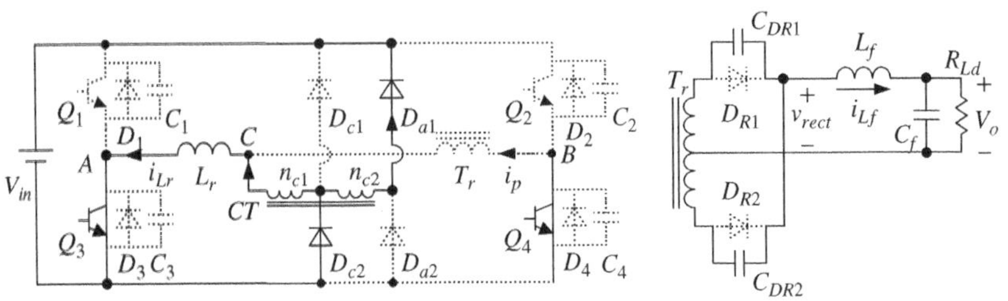

(f) $[t_4,t_5]$

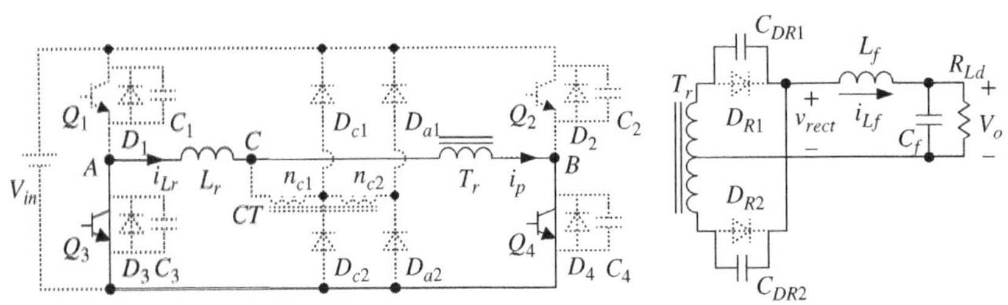

(g) $[t_5,t_6]$

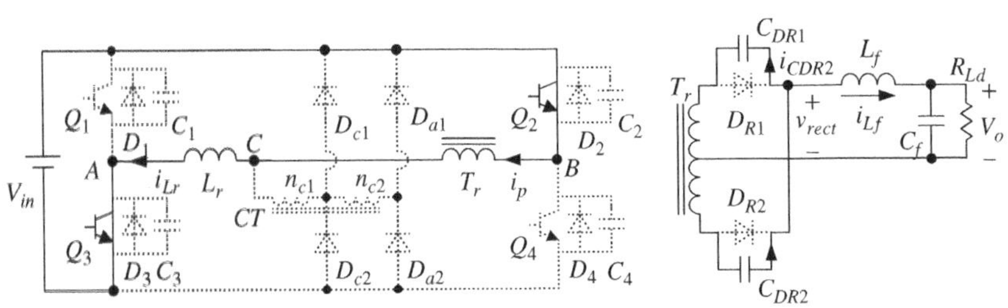

(h) $[t_6,t_7]$

Figure 7.14 *(Continued)*

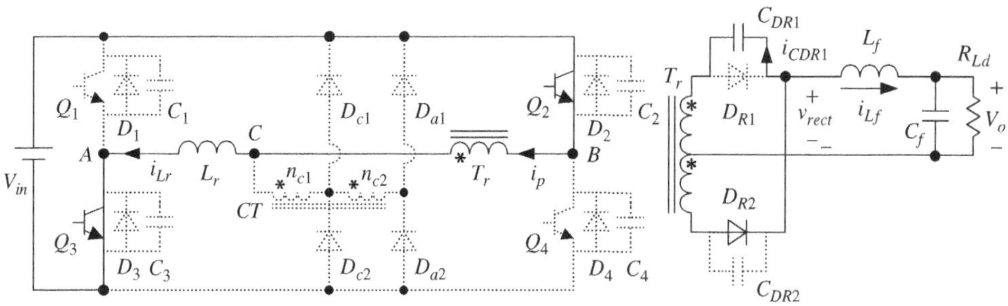

(i) $[t_7,t_8]$

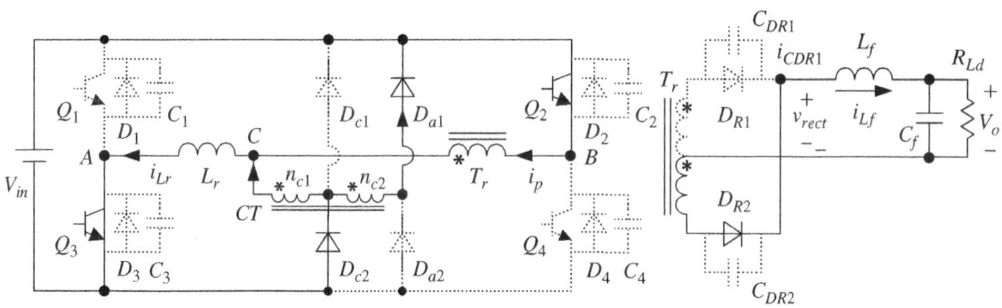

(j) $[t_8,t_9]$

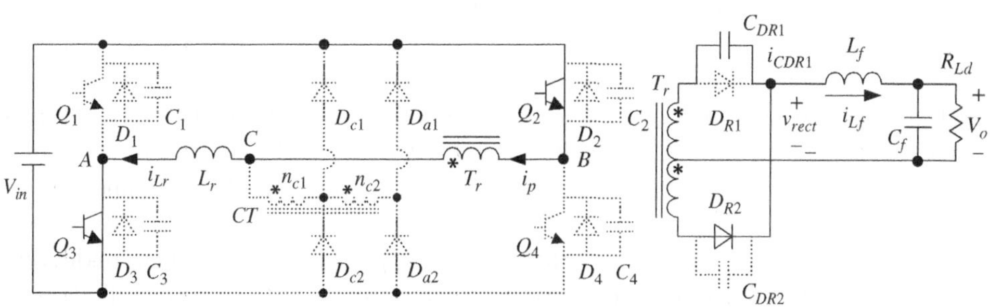

(k) $[t_9,t_{10}]$

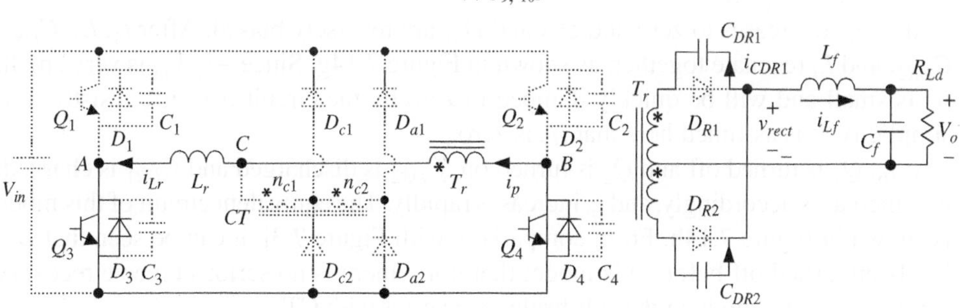

(l) $[t_1,t_3]$ (Case II)

Figure 7.14 *(Continued)*

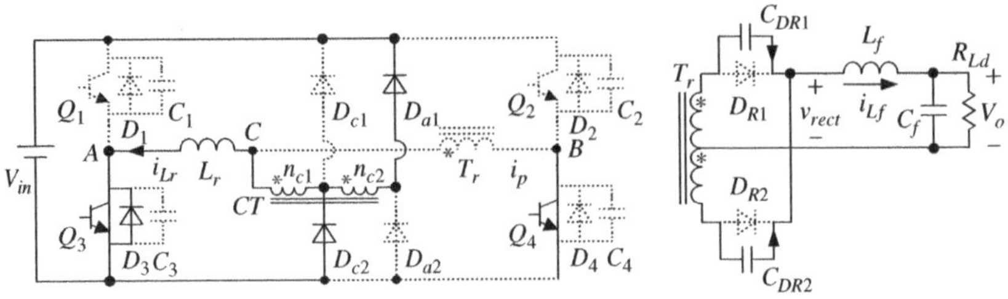

(m) $[t_3, t_4]$ (Case II)

Figure 7.14 (*Continued*)

7.4.2.1 Case I: $0.5V_{in}/Z_{r1} \le I_{Lf}(t_1)/K < V_{in}/Z_{r1}$ (Referring to Figure 7.13a)

If $0.5V_{in}/Z_{r1} \le I_{Lf}(t_1)/K < V_{in}/Z_{r1}$, i_p and i_{Lr} are negative when v_{CDR2} reduces to zero at t_1 and D_{R1} still conducts. After t_1, both D_{R1} and D_{R2} conduct, the voltages across the transformer primary and secondary windings are clamped at zero, and $v_{rect}=0$; thus, i_{Lf} begins to decline linearly, while i_{Lr} and i_p remain unchanged, as shown in Figure 7.14c. During this time, D_{R1} and D_{R2} begin to commute with the decrease in i_{Lf}, and i_{DR1} declines while i_{DR2} increases.

At t_2, i_{DR1} decreases to zero. C_{DR1} will then be charged and will resonate with the resonant inductor, as shown in Figure 7.14d. Accordingly, the voltage of node C will be lower than zero.

At t_3, the voltage of node C reduces to $-k_{ct}V_{in}$ and D_{c2} and D_{a1} conduct, as shown in Figure 7.14e. The induced CT primary winding voltage equals $-k_{ct}V_{in}$, which forces i_{Lr} to decrease rapidly. At the same time, $v_{rect}=k_{ct}V_{in}/K$. v_{rect} is very small because k_{ct} is designed to be far less than 1, so i_{Lf} continues to decrease and reaches zero at t_4.

After t_4, L_f resonates with C_{DR1} paralleling with C_{DR2} and D_{c2} continues conducting, as shown in Figure 7.14f. i_{Lr} continues decreasing, since $-k_{ct}V_{in}$ is still applied on the resonant inductor. i_p is zero. Since the voltage of node C is still $-k_{ct}V_{in}$, v_{CDR1} is always higher than v_{CDR2}, by $2k_{ct}V_{in}/K$.

At t_5, i_{Lr} decreases to zero and D_{c2} and D_{a1} are reversely biased. After t_5, L_r, C_{DR1}, C_{DR2}, and L_f resonate together, as shown in Figure 7.14g. Since $-k_{ct}V_{in}$ is very small, i_{Lr} is small and will be quickly damped to zero by the circuit parasitic resistor. For simplicity, it is assumed here that i_{Lr} is zero.

At t_6, Q_4 is turned off and Q_2 is turned on, C_{DR2} is discharged and C_{DR1} is charged, v_{rect} increases accordingly, and i_p increases rapidly. The equivalent circuit of this mode is shown in Figure 7.14h. From comparison with Figure 7.3j, it can be seen that D_{c2} has been turned off before this mode; therefore, there is no serious reverse recovery for the clamping diode in the full-bridge converter with CT.

At t_7, C_{DR2} is completely discharged and D_{R2} begins to conduct, C_{DR1} continues to be charged, and v_{rect} rises accordingly, as shown in Figure 7.14i.

At t_8, v_{rect} rises to $(V_{in}+k_{ct}V_{in})/K$ and the potential voltage of node C reduces to $-k_{ct}V_{in}$, so D_{c2} conducts. i_{Lr} reduces to i_p rapidly. The equivalent circuit of the mode is shown in Figure 7.14j.

At t_9, i_{Lr} equals i_p and D_{c2} turns off naturally. After t_9, L_r resonates with C_{DR1}. The maximum value of v_{rect} will never exceed $(1+k_{ct})V_{in}/K$.

7.4.2.2 Case II: $I_{Lf}(t_1)/K < 0.5V_{in}/Z_{r1}$ (Referring to Figure 7.13b)

If $I_{Lf}(t_1)/K < 0.5V_{in}/Z_{r1}$, v_{CDR2} is still positive when i_{DR1} reduces to zero at t_1; in other words, C_{DR2} is not completely discharged. The directions of i_{Lr} and i_p are changed at t_1.

After t_1, D_{R1} is turned off and C_{DR1} is charged, while C_{DR2} continues to be discharged. This is equivalent to the fact that L_r resonates with C_{DR1} and C_{DR2}. The voltage of node C decreases, as do v_{rect} and v_{CB}, while i_p and i_{Lr} continue to increase in the negative direction. The equivalent circuit of this mode is shown in Figure 7.14l. At t_2, the voltage of node C reduces to zero and i_{Lr} reaches its peak value of $-I_{Lr\_p}$. The voltage of node C then continues to decrease in the negative direction and i_{Lr} begins to increase.

At t_3, the voltage of node C reduces to $-k_{ct}V_{in}$ and D_{c2} conducts. i_{Dc2} flows into the synonym end of the CT primary winding and the dot side of the CT secondary winding, forcing D_{a1} to conduct. V_{in} is applied on the CT secondary winding and the induced voltage across the CT primary winding is $-k_{ct}V_{in}$; this is applied on L_r, forcing i_{Lr} to decrease quickly. v_{BC} is clamped at $k_{ct}V_{in}$ by CT and D_{c2}, and $v_{rect}=k_{ct}V_{in}/K$. i_{Lf} decreases linearly due to the smaller v_{rect}, and i_{Lf} reduces to zero at t_4. During this time interval, both C_{DR1} and C_{DR2} are discharged; they have the same discharged currents, which share i_{Lf}, so i_p is kept at zero. The equivalent circuit of this mode is shown in Figure 7.14m.

During $[t_4, t_{10}]$, the operation of the converter is similar to that in case I. It will not be repeated here.

From this analysis, it can be seen that no matter under heavy or light load conditions, the reset voltage source will be induced so long as there is current flowing through the clamping diodes. This means the clamping diode current declines rapidly, significantly reducing its conduction time and thus the conduction losses on the clamping diodes, leading-leg switches, and resonant inductor. This avoids the reverse recovery of the clamping diodes.

7.5 Choice of Current Transformer Winding-Turns Ratio

7.5.1 Clamping Diode Current-Reset Time

Generally, k_{ct} is designed to be far less than 1. Hence, according to Equation 7.24, the peak current of the clamping diodes under heavy load conditions can be obtained as:

$$I_{Dc\max\_h} \approx V_{in}/Z_{r1} \tag{7.26}$$

The conduction voltage drops of the switches and clamping diodes are much lower than the reset voltage and are thus neglected; that is:

$$L_r \frac{di_{Lr}}{dt} = k_{ct} V_{in} \tag{7.27}$$

Assuming that the output filter inductance is quite large, and neglecting its current ripple, the resetting time of the clamping diode current under heavy load conditions can be obtained by Equations 7.26 and 7.27 as:

$$\Delta t_{\_h} = \frac{1}{k_{ct}\omega_1} \tag{7.28}$$

According to the analysis in Section 7.2, under light load conditions the smaller the load current, the larger the peak value of the resonant inductor current; that is, the larger the peak value of the clamping diode current. From Equation 7.18, the maximum value of the clamping diode current is:

$$I_{Dc\max\_l} = \frac{V_{in}}{\sqrt{2}Z_{r1}} \tag{7.29}$$

Combining Equations 7.27 and 7.29, the maximum clamping diode conduction time under light load conditions is:

$$\Delta t_{\_l\max} = \frac{1}{\sqrt{2}k_{ct}\omega_1} \tag{7.30}$$

It can be seen from Figure 7.13 that the maximum clamping diode conduction time must be less than the time interval of the zero state under light load conditions; that is:

$$\Delta t_{\_l\max} < T_s(1 - D_p)/2 \tag{7.31}$$

where D_p is the primary duty cycle of the full-bridge converter.

7.5.2 Output Rectifier Diode Voltage Stress

According to the analysis in Section 7.4, the voltage stress of the output rectifier diode in the full-bridge converter with CT is $2(V_{in} + k_{ct}V_{in})/K$, which is slightly higher than the full-bridge converter without CT. The difference is:

$$\Delta V_{DR} = \frac{2(V_{in} + k_{ct}V_{in})}{K} - \frac{2V_{in}}{K} = \frac{2k_{ct}V_{in}}{K} \tag{7.32}$$

7.5.3 Current Transformer Winding-Turns Ratio

The conduction time of the clamping diode should ideally be as short as possible, in order to reduce conduction loss. From Equations 7.28 and 7.30, k_{ct} should therefore be as large as possible. However, from Equation 7.32, a smaller k_{ct} is helpful in reducing the voltage stress of the output rectifier diode. The choice of k_{ct} is thus a trade-off.

Generally, the primary duty cycle D_p is relatively smaller under light load conditions. Assuming that the maximum clamping diode conduction time under light load conditions is 10% of the switching period, we can get:

$$\Delta t_{\_l\max} = \frac{1}{\sqrt{2}k_{ct}\omega_1} < \frac{T_s}{10} \tag{7.33}$$

Rewriting Equation 7.33, the turns ratio of CT is given by:

$$k_{ct} > \frac{5\sqrt{2}}{T_s\omega_1} \tag{7.34}$$

7.6 Experimental Verification

This section presents an experimental evaluation of a 1 kW prototype of the ZVS PWM full-bridge converter with CT. The main specifications are:

- input voltage $V_{in} = 270\,\text{V} \pm 10\%$;
- output voltage $V_o = 54\,\text{V}$; and
- maximum output current $I_o = 20\,\text{A}$.

The main power devices and components are:

- $Q_1 \sim Q_4$: IRFP450;
- output rectifier diodes (center-tapped rectifier): DSEP 30-03A;
- clamping diode D_{c1} and D_{c2} DSEI30-06A;
- resonant inductor $L_r = 9\,\mu\text{H}$;
- transformer turns ratio $K = 15:4$;
- output filter inductor $L_f = 23\,\mu\text{H}$;
- output filter capacitor $C_f = 560\,\mu\text{F} \times 2$;
- CT turns ratio $k_{ct} = n_{c1}:n_{c2} = 6:77$;
- auxiliary diode D_{a1} and D_{a2}: BYV26C; and
- switching frequency: 100 kHz.

Figure 7.15 shows the waveforms of the v_{AB}, i_p, i_{Lr}, i_{Dc1}, i_{Dc2}, and v_{rect} (from top to bottom) of the ZVS PWM full-bridge converters with and without CT at full load under the nominal input voltage of 270 V. It can be seen that the clamping diode conduction time is much shorter when CT is introduced to the full-bridge converter and that v_{rect} is slightly higher with CT than without, which is well in agreement with the previously mentioned theoretical analysis.

Figure 7.16 shows the waveforms of the full-bridge converters with and without CT at light load ($I_o = 30\,\text{mA}$) under the nominal input voltage of 270 V. It can be seen that

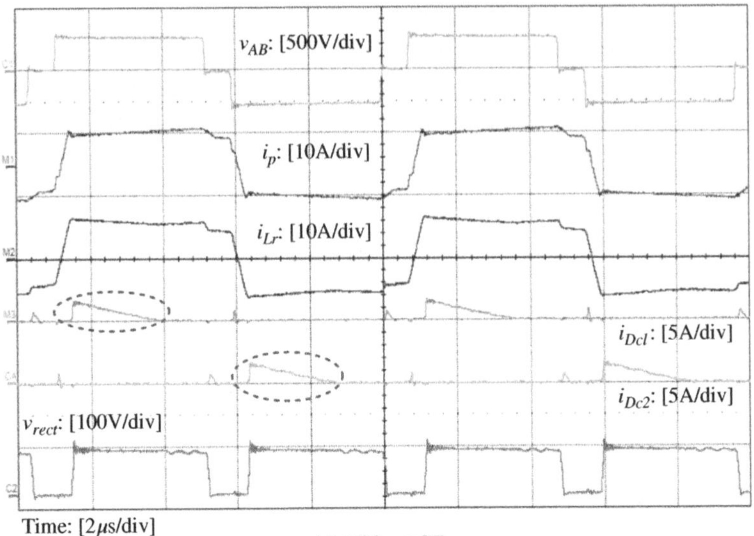

(a) Without CT

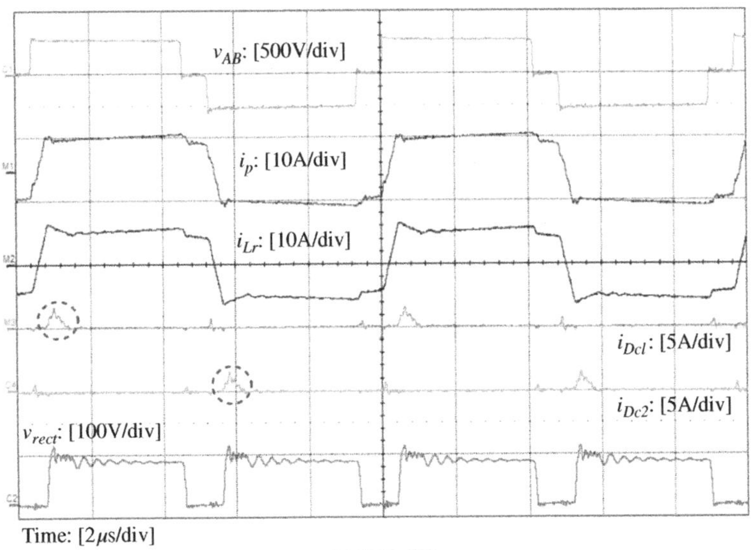

(b) With CT

Figure 7.15 Waveform comparison of a full-bridge converter with and without CT at full load ($I_o = 20$ A)

the two clamping diodes conduct alternately for nearly half of the switching period in the full-bridge converter without CT. When the full-bridge converter transits from zero state ($v_{AB} = 0$) to active state ($v_{AB} = +V_{in}$ or $-V_{in}$), the clamping diodes are turned off with reverse recovery; they are apt to be destroyed. For the full-bridge converter with CT, the clamping diode current decays rapidly without reverse recovery and the

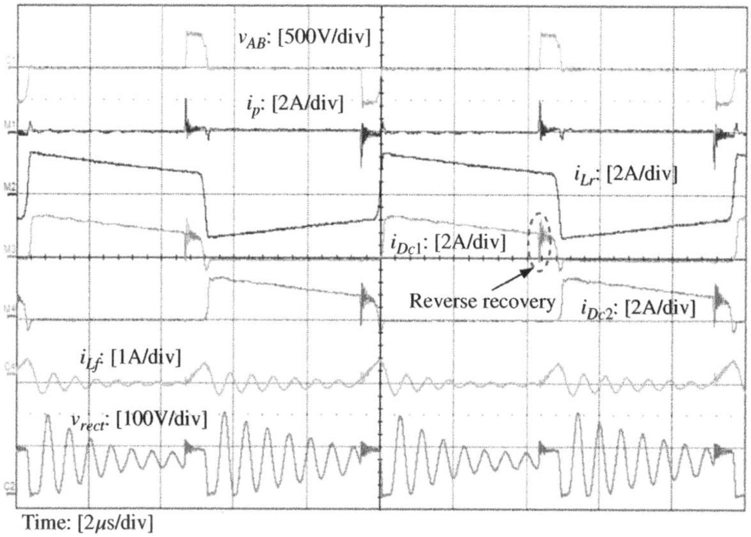

(a) Without CT

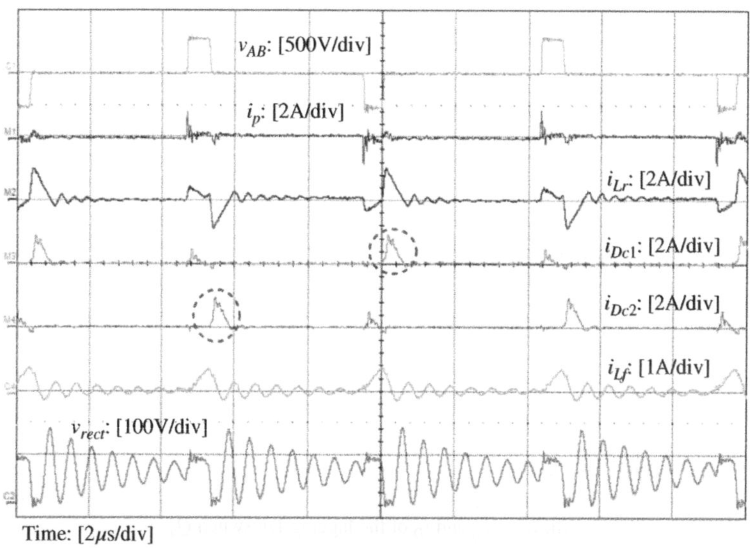

(b) With CT

Figure 7.16 Waveform comparison of a full-bridge converter without and with CT at light load ($I_o = 30\,\text{mA}$)

conduction time is shortened thanks to the induced reset voltage when the full-bridge converter transits from active to zero state.

Figure 7.17 shows the waveforms of the gate-source voltage v_{GS}, the drain-source voltage v_{DS}, and the drain current i_D of the leading-leg switch Q_1 and lagging-leg switch Q_4, respectively. It can be seen that all of the switches realize ZVS.

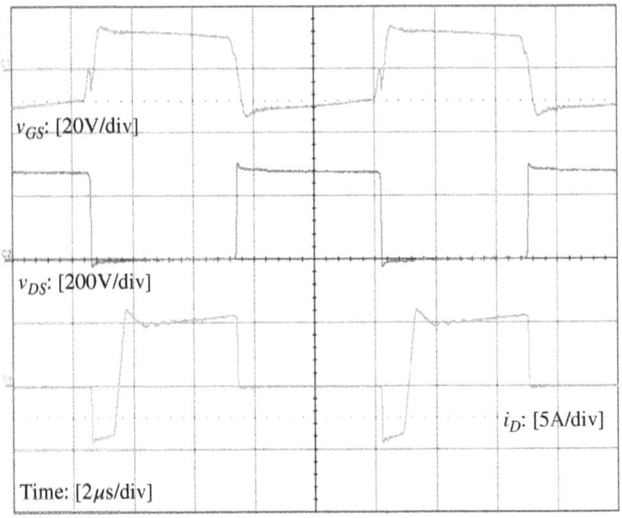

(a) v_{GS}, v_{DS} and i_D of the leading-leg switch Q_1

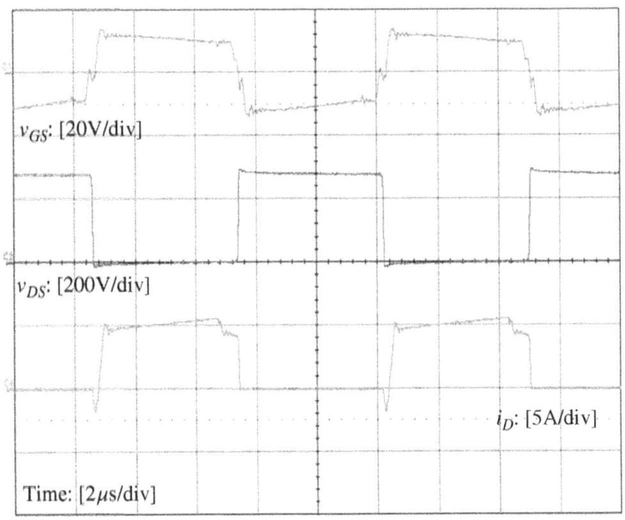

(b) v_{GS}, v_{DS} and i_D of the lagging-leg switch Q_4

Figure 7.17 Waveforms of the leading and the lagging switch

Figure 7.18 shows conversion efficiency comparisons of the full-bridge converters with and without CT. Figure 7.18a shows efficiency versus output current under a normal input voltage of 270 V. Figure 7.18b shows efficiency at full load versus input voltage. It can be seen that when the CT is incorporated, the efficiency of the full-bridge converter is improved, due to the reduced conduction loss of the clamping diodes, the leading-leg switches, and the resonant inductor.

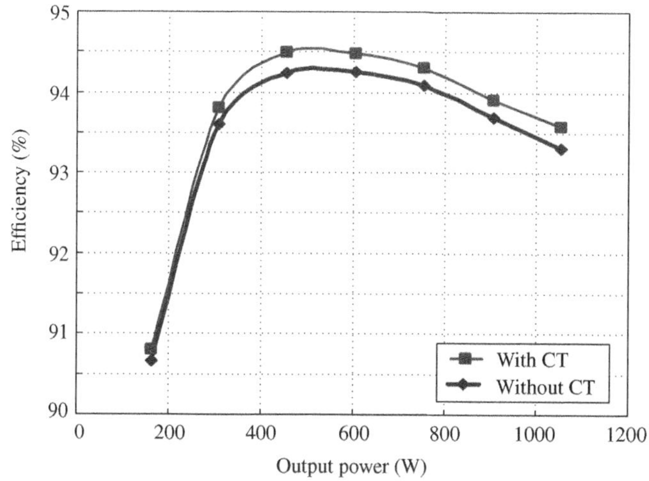

(a) Efficiency versus output current under normal input voltage

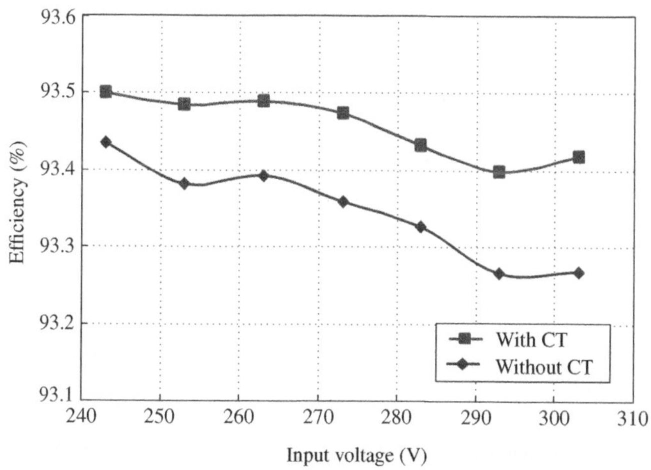

(b) Efficiency at full load versus input voltage

Figure 7.18 Efficiency comparison of full-bridge converters with and without CT

7.7 Summary

This chapter described the introduction of a CT-based clamping diode current-reset scheme into a ZVS PWM full-bridge converter with two clamping diodes. This converter is designed to overcome the hard-turn-off issue of the clamping diode under light load conditions. The proposed converter can still eliminate the voltage oscillation resulting from the reverse recovery of the rectifier diodes, while also offering the following additional advantages over the traditional ZVS PWM full-bridge converter:

1. The clamping diode current can be reduced to zero rapidly over the full load range. Thus the conduction loss of the clamping diodes, leading-leg switches, and resonant inductor are reduced and the conversion efficiency can be increased.
2. The clamping diodes can be turned off naturally, without reverse recovery, over the whole input voltage and load range. The reliability of the full-bridge converter is thus improved.

The operating principle of the improved full-bridge converter has been analyzed in detail and validated by a 1 kW prototype.

References

1. Redl, R., Sokal, N.O. and Balogh, L. (1991) A novel soft-switching full-bridge dc/dc converter: analysis, design considerations and experimental results at 1.5 kW, 100 kHz. *IEEE Trans. Power Electron.*, **26** (3), 408–418.
2. Redl, R., Balogh, L. and Edwards, D.W. (1993) Switch transitions in the soft-switching full-bridge PWM phase-shift dc/dc converter: analysis and improvements. Proceedings of the International Telecommunication Energy Conference, pp. 350–357.
3. Redl, R., Balogh, L. and Edwards, D.W. (1994) Optimum ZVS full-bridge dc/dc converter with PWM phase-shifted control: analysis, design considerations, and experimental results. Proceedings of the IEEE Applied Power Electronics Conference and Exposition, pp. 159–165.
4. Ruan, X. and Liu, F. (2004) An improved ZVS PWM full-bridge converter with clamping diodes. Proceedings of the IEEE Power Electronics Specialists Conference, pp. 1476–1481.
5. Yin, L., Chen, Q., Peng, B. *et al.* (2007) Key issues of clamping diodes in DCM phase-shift full-bridge converter. Proceedings of the IEEE Power Electronics Specialists Conference, pp. 1721–1725.
6. Chen, W., Ruan, X., Chen, Q. and Ge, J. (2010) Zero-voltage-switching PWM full-bridge converter employing auxiliary transformer to reset the clamping diode current. *IEEE Trans. Power Electron.*, **25** (5), 1149–1162.
7. Chen, W., Ruan, X. and Zhang, R. (2008) A novel zero-voltage-switching PWM full bridge converter. *IEEE Trans. Power Electron.*, **23** (2), 793–801.
8. Chen, Q., Yin, L., Wang, J. *et al.* (2008) Phase-shifted full-bridge PWM converter with clamping diodes and current transformer. Proceedings of the IEEE Power Electronics Specialists Conference, pp. 2403–2409.

8

Zero-Voltage-Switching PWM Full-Bridge Converters with Current-Doubler Rectifiers

Chapter 3 presented the operating principle of the basic zero-voltage-switching (ZVS) pulse-width-modulation (PWM) full-bridge converter. The lagging leg can only utilize the energy stored in the transformer leakage inductor to achieve ZVS, and ZVS is lost at light load. An external resonant inductor is always introduced to connect in series with the transformer primary winding, in order to widen the load range for ZVS. However, during the commutation of the output rectifier diodes, the resonant inductor resonates with the junction capacitor of the outgoing rectifier diode, leading to voltage oscillation and voltage spike across the outgoing rectifier diode. Two clamping diodes can be introduced to the primary side to clamp both the primary and the secondary voltages, thus eliminating the voltage oscillation across the outgoing rectifier diode, as discussed in Chapter 6. Further, as presented in Chapter 7, a current transformer resetting circuit is added to reset the clamping diode current. This not only improves the efficiency, due to regeneration of the surplus energy stored in the resonant inductor, but also avoids the hard turn-off of the clamping diodes at light load. Although the resonant inductor can widen the load range of ZVS, it cannot make it too large due to the resultant duty cycle loss. That is to say, when the load is lighter, the lagging leg still loses ZVS.

Chapter 1 presented the operating principle of the full-bridge converter with current-doubler rectifier (CDR). The output filter inductor current flows in the negative direction at light load, and this negative current is reflected to the primary side in zero state, forcing the primary current to increase. By utilizing this characteristic, reference [1] proposes a CDR ZVS PWM full-bridge converter. The energy stored in the output filter inductors is used to achieve ZVS for both the leading and the

Soft-Switching PWM Full-Bridge Converters: Topologies, Control, and Design, First Edition. Xinbo Ruan.

lagging leg over a wide load range, while the voltage oscillation and voltage spike across the output rectifier diodes resulting from the reverse recovery is avoided. In order to ensure the negative current of the output filter inductor is reflected to the primary side, the primary current at zero state should decay rapidly to the output filter inductor current reflected to the primary side. It is only the conduction voltage drop of the power switches that forces the primary current to decay, and this is very small, so the leakage inductor must be very small, which requires a special manufacturing method to produce the transformer.

Based on the CDR ZVS PWM full-bridge converter [1], a blocking capacitor can be introduced in series with the primary side [2, 3]. As the voltage of the blocking capacitor is quite large compared to the conduction voltage drop of the power switches, the primary current can decay rapidly even if the leakage inductance is relatively large. Therefore, there is no special limit for the leakage inductance. This chapter will analyze the operating principle of the improved CDR ZVS PWM full-bridge converter and discuss the realization of ZVS for the leading and the lagging leg. A simplified design procedure for the improved converter is given, along with detailed experimental results from a 540 W output prototype, which verify the operating principle of the improved converter.

8.1 Operating Principle

The CDR ZVS PWM full-bridge converter is shown in Figure 8.1a, where Q_1 to Q_4 are power switches, D_1 to D_4 are body diodes of Q_1 to Q_4, and C_1 to C_4 are intrinsic capacitors of Q_1 to Q_4. L_{lk} is the leakage inductor of the transformer and C_b is the blocking capacitor. The rectifier diodes D_{R1} and D_{R2} form a CDR with the output filter inductors L_{f1} and L_{f2}. C_f is the output filter capacitor and R_{Ld} is the load. The converter employs the popular phase-shift control. Q_1 and Q_3 form the leading leg and Q_2 and Q_4 the lagging.

This section analyzes the operating principle of the improved ZVS PWM full-bridge converter. The following assumptions are made:

1. All the switches and diodes are ideal.
2. All the inductors and capacitors are ideal.
3. $C_1 = C_3 = C_{lead}$ and $C_2 = C_4 = C_{lag}$.
4. $L_{f1} = L_{f2} = L_f$.
5. C_f is large enough to be treated as a voltage source V_o, which is the output voltage.

There are 12 switching modes in a switching period. Since the first half-period is similar to the second, only the first is given here for brevity, which includes 6 switching modes and 1 initial switching mode. Figure 8.2 shows the equivalent circuits of each.

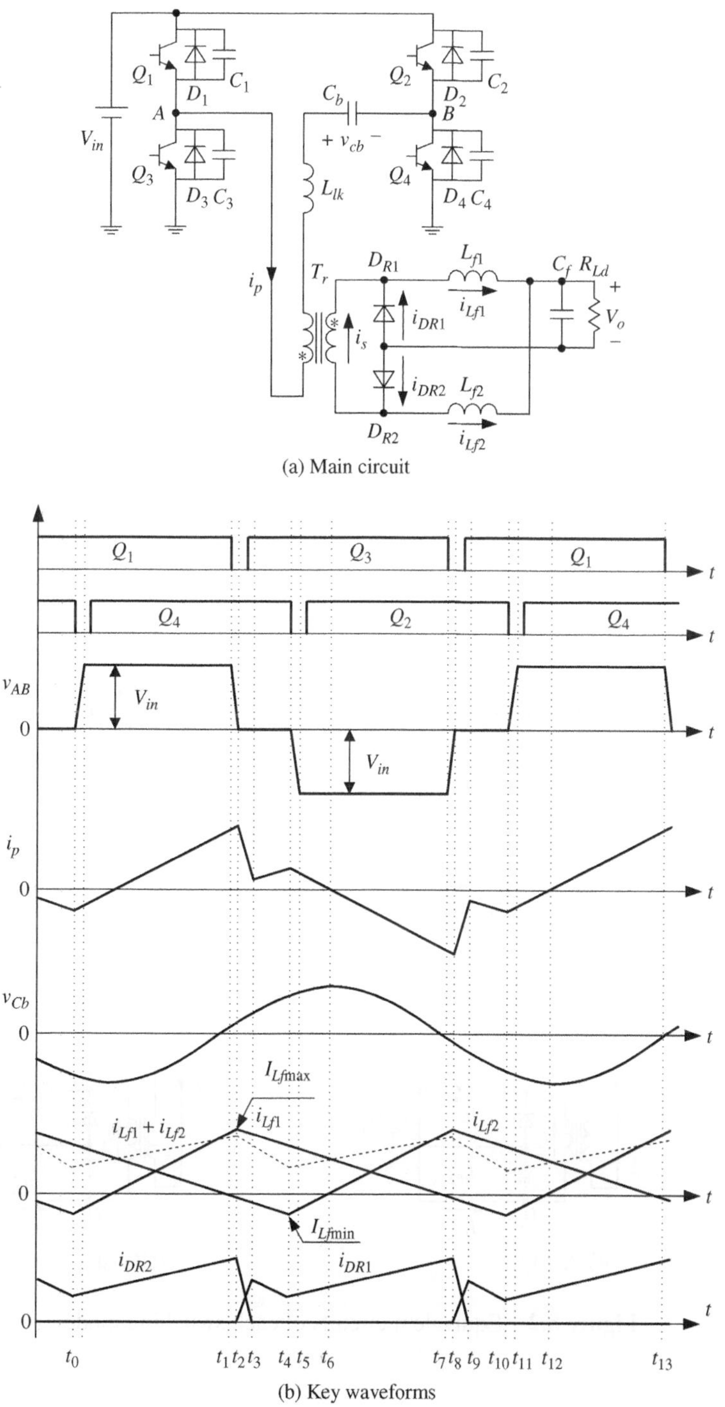

Figure 8.1 Improved CDR ZVS PWM full-bridge converter

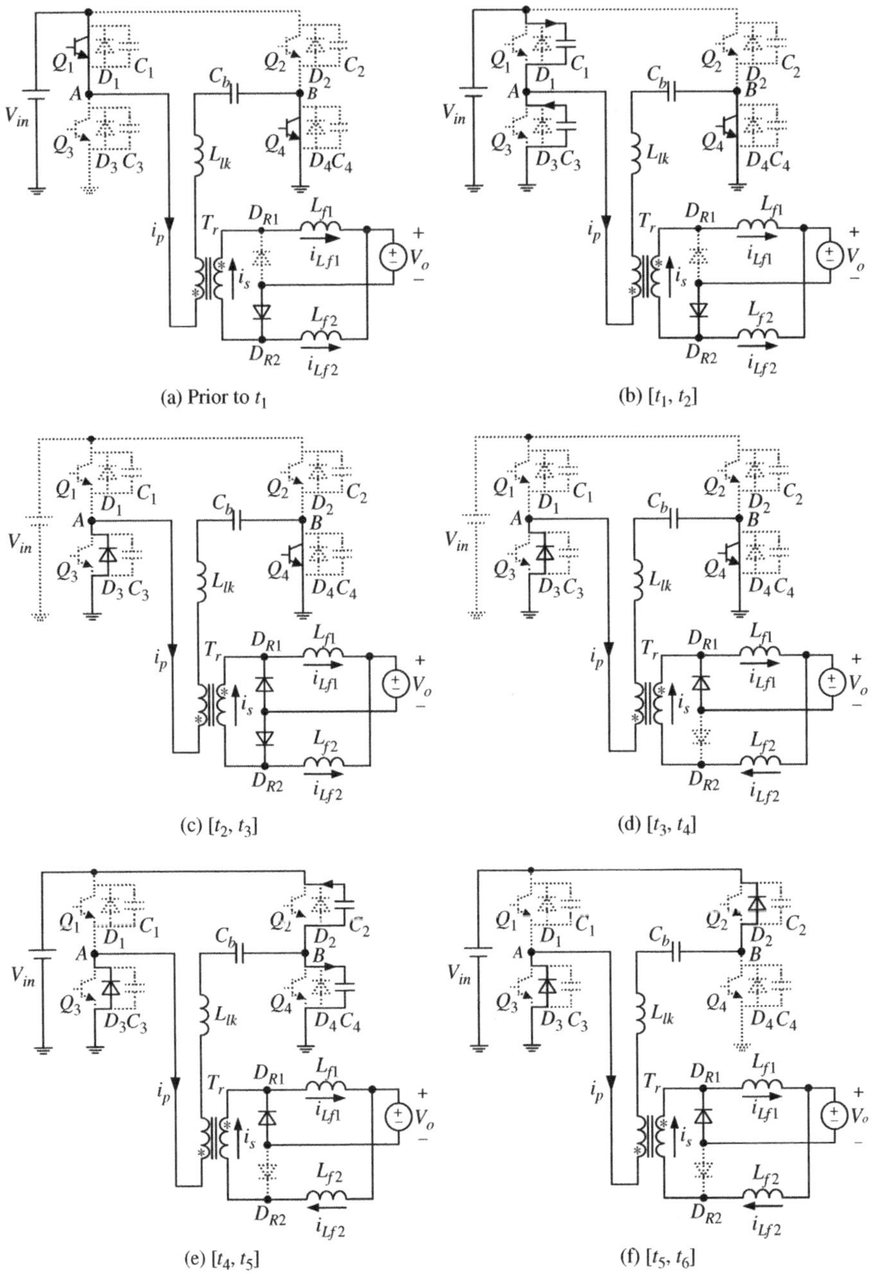

Figure 8.2 Equivalent circuits of each switching mode

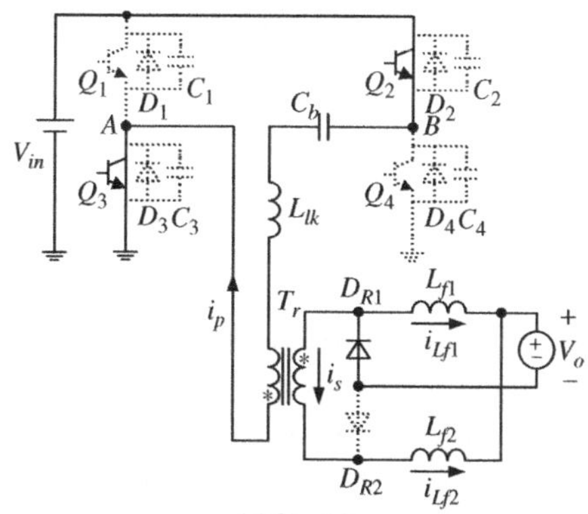

(g) $[t_6, t_7]$

Figure 8.2 (*Continued*)

1. **Mode 0, prior to t_1 (Figure 8.2a):** Prior to t_0, Q_1 and Q_4 are conducting and the primary current i_p flows through Q_1, the primary winding, the blocking capacitor, and Q_4. D_{R2} is conducting and D_{R1} is off. The two output filter inductor currents and the primary current are expressed as:

$$i_{Lf1}(t) = I_{10} + \frac{\frac{V_{in}}{K} - V_o}{L_f}(t - t_0) \tag{8.1}$$

$$i_{Lf2}(t) = I_{20} - \frac{V_o}{L_f}(t - t_0) \tag{8.2}$$

$$i_p(t) = i_{Lf1}(t)/K \tag{8.3}$$

where K is the turns ratio of the primary to secondary windings of the transformer and I_{10} and I_{20} are the currents of the two filter inductors at t_0.

2. **Mode 1, $[t_1, t_2]$ (Figure 8.2b):** Q_1 is turned off at t_0. i_p charges C_1 and discharges C_3. As C_1 and C_3 limit the rising slope of the voltage across Q_1, Q_1 is turned off with zero voltage. At the same time, i_p also charges C_b. $i_p = i_{Lf1}/K$, and L_{f1} is large enough that i_{Lf1} stays nearly constant in this mode, so i_p too stays nearly constant. Therefore, the voltage of C_1 increases linearly from zero and the voltage of C_3 decreases linearly from V_{in}, expressed as:

$$v_{C1}(t) = \frac{I_p(t_1)}{2C_{lead}}(t - t_1) \tag{8.4}$$

$$v_{C3}(t) = V_{in} - \frac{I_p(t_1)}{2C_{lead}}(t - t_1) \tag{8.5}$$

At t_2, the voltage of C_3 decays to zero and the body diode D_3 conducts naturally. The time period of mode 1 is:

$$t_{12} = \frac{2C_{lead}V_{in}}{I_p(t_1)} \tag{8.6}$$

3. **Mode 2, [t_2, t_3] (Figure 8.2c):** D_3 is conducting, clamping the voltage across Q_3 at zero, so Q_3 can be zero-voltage turn-on. The delay time between the gate drives of Q_3 and of Q_1 should be greater than the interval of mode 1; that is, $t_{d(lead)} > t_{12}$. During this interval, $v_{AB}=0$, v_{Cb} forces the primary current i_p to decrease, and the secondary current i_s decreases correspondingly, which makes D_{R1} conduct. As the two rectifier diodes conduct simultaneously, both the primary and secondary voltages of the transformer are clamped at zero. Thus, v_{Cb} is fully applied to L_{lk}, and C_b resonates with the latter. Moreover, $-V_o$ is applied on the two output filter inductors, causing i_{Lf1} and i_{Lf2} to decrease linearly. The expressions of the two output filter inductor currents, the primary current, and the blocking capacitor voltage are:

$$i_{Lf1}(t) = I_{Lf1}(t_2) - \frac{V_o}{L_f}(t - t_2) \tag{8.7}$$

$$i_{Lf2}(t) = I_{Lf2}(t_2) - \frac{V_o}{L_f}(t - t_2) \tag{8.8}$$

$$i_p(t) = -\frac{V_{Cb}(t_2)}{\omega L_{lk}} \sin \omega(t - t_2) + I_p(t_2) \cos \omega(t - t_2) \tag{8.9}$$

$$v_{Cb}(t) = \omega L_{lk} I_p(t_2) \sin \omega(t - t_2) + V_{Cb}(t_2) \cos \omega(t - t_2) \tag{8.10}$$

where $\omega = 1/\sqrt{L_{lk}C_b}$.

Prior to this mode, or while in it, i_{Lf2} decreases to become negative. At t_3, i_s decreases to be equal to $-i_{Lf2}$, and $i_{DR2}=0$, so D_{R2} turns off naturally and D_{R1} continues conducting and carries both of the output filter inductor currents.

4. **Mode 3, [t_3, t_4] (Figure 8.2d):** D_3 and Q_4 continue conducting and $v_{AB}=0$. v_{Cb} is sufficiently small compared with the reflected output voltage that it can be neglected. Thus, the voltages applied on the two output filter inductors are approximately $-V_o$. In this case, i_{Lf1} and i_{Lf2} continue to decrease linearly. During the interval, $i_s=-i_{Lf2}$, so $i_p=-i_{Lf2}/K$. Because i_{Lf2} decreases and is negative, i_p is forced to increase.
5. **Mode 4, [t_4, t_5] (Figure 8.2e):** At t_4, Q_4 is turned off and i_p charges C_4 and discharges C_2. Q_4 is zero-voltage turn-off thanks to C_2 and C_4. At the same time, i_p charges C_b. As i_p is the reflected output filter inductor current i_{Lf2} to the primary side and L_{f2} is large enough to be treated as a constant current source, i_p remains

nearly constant. Therefore, the voltage of C_4 increases linearly from zero and the voltage of C_2 decreases linearly from V_{in}, expressed as:

$$v_{C4}(t) = \frac{I_p(t_4)}{2C_{lag}}(t - t_4) \tag{8.11}$$

$$v_{C2}(t) = V_{in} - \frac{I_p(t_4)}{2C_{lag}}(t - t_4) \tag{8.12}$$

The voltage of C_2 decreases to zero at t_5 and the body diode D_2 conducts naturally. The time period of mode 5 is:

$$t_{45} = \frac{2C_{lag}V_{in}}{I_p(t_4)} \tag{8.13}$$

6. **Mode 5, [t_5, t_6] (Figure 8.2f):** As D_2 conducts, Q_2 can be turned on with zero voltage. The delay time between the gate drive signals of Q_2 and of Q_4 should be greater than the time period of mode 5; that is, $t_{d(lag)} > t_{45}$. During this interval, $v_{AB} = -V_{in}$, L_{lk} is quite small, and the voltage of C_b is very small compared with V_{in} (which can be neglected), so i_{Lf1} decreases and i_{Lf2} increases linearly. Because $i_p = -i_{Lf2}/K$, i_p decreases linearly. The voltage of C_b continues increasing.

   At t_6, i_p decreases to zero, D_2 and D_3 turn off naturally, and the voltage of C_b reaches its maximum value.
7. **Mode 6, [t_6, t_7] (Figure 8.2g):** Q_2 and Q_3 conduct, i_{Lf1} decreases, i_{Lf2} increases, and i_p increases in the negative direction. The voltage of C_b begins to decrease.

   At t_7, Q_3 is turned off, beginning the second half-cycle [t_7, t_{13}], which is similar to the first [t_1, t_7].

8.2 Realization of ZVS for the Switches

From the analysis in Section 8.1, it can be seen that the leading leg realizes ZVS through the use of the energy stored in the output filter inductor when the current is at its maximum (at t_1 or t_7), while the lagging leg realizes ZVS through the use of the energy stored in the output filter inductor when the current is at its minimum (negative at t_4 or t_{10}).

As can be seen from Figure 8.1b, the average current of the output filter inductor is:

$$I_{Lfavg} = \frac{I_{Lf\max} + I_{Lf\min}}{2} = \frac{I_o}{2} \tag{8.14}$$

where $I_{Lf\max}$ and $I_{Lf\min}$ are the maximum and minimum current of the output filter inductor, respectively, and I_o is the load current.

During the interval $[t_2, t_{10}]$, I_{Lf1} decays linearly from $I_{Lf\max}$ to $I_{Lf\min}$; that is:

$$I_{Lf\max} - I_{Lf\min} = \frac{V_o}{L_f}\left(1 - \frac{D_y}{2}\right)T_s \tag{8.15}$$

where T_s is the switching period and D_y is the duty cycle of the converter, represented by $D_y = (t_1 - t_0)/\frac{T_s}{2}$.

Solving Equations 8.14 and 8.15 gives:

$$I_{Lf\max} = \frac{I_o}{2} + \frac{V_o(2 - D_y)T_s}{4L_f} \tag{8.16}$$

$$I_{Lf\min} = \frac{I_o}{2} - \frac{V_o(2 - D_y)T_s}{4L_f} \tag{8.17}$$

Note that $I_{Lf\min}$ is a negative value.

From Equations 8.16 and 8.17, we observe the following: (i) the larger the output current, the larger $I_{Lf\max}$ and the smaller $|I_{Lf\min}|$; thus, it is easier to realize ZVS in the leading leg at heavy load than at light, whereas it is easier to realize ZVS in the lagging leg at light load than at heavy; and (ii) because $I_{Lf\max} > |I_{Lf\min}|$, it is easier to realize ZVS in the leading leg than in the lagging, if the intrinsic capacitors and the delay time of the leading and the lagging leg are equal. The worst case is thus to achieve ZVS for the lagging switch at full load.

As analyzed in Section 8.1, in order to realize ZVS for the lagging leg through the use of the energy of the output filter inductor, two conditions should be satisfied during zero state ($v_{AB} = 0$): (i) the output filter inductor current should become negative; and (ii) the primary current should decay rapidly to reach the reflected output filter inductor current, although it should be positive beforehand. The two conditions can be satisfied through proper design of the output filter inductor and blocking capacitor, as will be explained in the next section.

Fortunately, when the two conditions are satisfied, the output rectifier diodes complete the commutation naturally during zero state, prior to the switching of the primary voltage v_{AB} to the active mode ($v_{AB} = +V_{in}$ or $-V_{in}$). This eliminates the reverse recovery of the rectifier diodes, and no oscillation occurs.

8.3 Design Considerations

This section discusses the design considerations for the improved CDR ZVS PWM full-bridge converter, especially the design of the output filter inductance and the blocking capacitor.

The input specifications are as follows:

- input voltage $V_{in} = 250$ Vdc $\pm 20\%$;
- output voltage $V_o = 54$ Vdc;
- output current $I_o = 10$ A;
- switching frequency $f_s = 100$ kHz; and
- leakage inductance measured at the switching frequency $L_{lk} = 0.46\ \mu$H.

8.3.1 *Transformer Winding-Turns Ratio*

The relationship between the output and the input voltage of the ZVS PWM full-bridge converter with CDR in continuing current mode (CCM) is:

$$D_y = \frac{V_o}{V_{in}/(2K)} = \frac{2KV_o}{V_{in}} \tag{8.18}$$

Here, CCM means that the sum of the two filter inductor currents is greater than zero (i.e., $i_{Lf1} + i_{Lf2} > 0$) when $v_{AB} = 0$, as shown in Figure 8.1b.

Rewriting Equation 8.18, the turns ratio of the transformer is derived as:

$$K = \frac{D_y V_{in}}{2V_o} \tag{8.19}$$

If $D_{y\max} = 0.8$ at the lowest input voltage $V_{in\min} = 200$ V then $K = 1.48$. We choose $K = 1.5$.

8.3.2 *Output Filter Inductance*

It is better for a filter inductor to be larger in order to reduce the current ripple. However, in order to realize ZVS for the lagging leg at full load, it should be quite small so that the filter inductor current can flow in the negative direction. We should thus determine the maximum value of the filter inductor to ensure ZVS for the lagging leg at full load.

As can be seen in Figure 8.1b, at t_4 $i_{Lf} = -I_{Lf\min}/K$. Substituting this into Equation 8.13 yields:

$$t_{45} = \frac{2C_{lag}V_{in}}{I_p(t_4)} = \frac{2C_{lag}V_{in}}{-I_{Lf\min}/K} \tag{8.20}$$

According to Equations 8.17, 8.18, and 8.20, the maximum value of the output filter inductor is derived as:

$$L_{f\max} = \frac{t_{45}V_o(V_{in} - KV_o)}{\left(4KC_{lag}V_{in}^2 + t_{45}V_{in}I_{o\max}\right)f_s} \tag{8.21}$$

Equation 8.21 shows that L_f is determined by V_{in} and t_{45}. As I_o increases, $|I_{Lf\min}|$ decreases and t_{45} increases. In other words, t_{45} reaches its maximum value at full load. In order to reduce the turn-off loss of the lagging-leg switch at light load, we can choose the delay time of the gate drives $t_d = t_{45\max} = 7 \cdot t_f$, where t_f is the turn-off time of the lagging-leg switch. Here IRF450 is selected as the switch with $C_{oss} = 720\,\text{pF}$ when its drain-to-source voltage $V_{ds} = 25\,\text{V}$ and $t_f = 44\,\text{ns}$. C_{oss} is nonlinear and is inversely proportional to the square root of the voltage applied on the drain-to-source; that is:

$$C_{oss} = C_o' / \sqrt{V_{ds}} \tag{8.22}$$

where C_o' is a constant that depends on the construction of the device [4]. Hence, the expression of C_{oss} is:

$$C_{oss} = 720 \times 10^{-12} \times \sqrt{25/V_{ds}} \tag{8.23}$$

For simplification of analysis, the intrinsic capacitor can be replaced by a constant value, which is called an effective capacitor. The effective capacitance is C_{oss} multiplied by 4/3, and V_{ds} is replaced by V_{in} [4]. Therefore, C_{lead} and C_{lag} can be expressed by:

$$C_{lead} = C_{lag} = \frac{4}{3} \times 720 \times 10^{-12} \times \sqrt{25/V_{in}} \tag{8.24}$$

Substituting Equation 8.24 into Equation 8.21, the maximum value of the output filter inductor under different input voltages can be depicted as shown in Figure 8.3, from which we determine $L_f = 28\,\mu\text{H}$.

As L_f is determined, we can calculate $I_{Lf\max}$ and $I_{Lf\min}$. When the full-bridge converter with CDR works in CCM, substituting Equation 8.18 into Equations 8.16

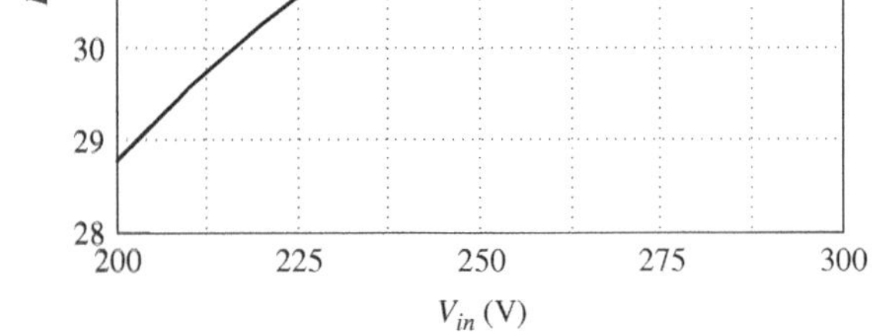

Figure 8.3 Maximum filter inductance versus input voltage

and 8.17, $I_{Lf\max}$ and $I_{Lf\min}$ can be derived as:

$$I_{Lf\max\_CCM} = \frac{I_o}{2} + \frac{V_o(V_{in} - KV_o)T_s}{2V_{in}L_f} \tag{8.25}$$

$$I_{Lf\min\_CCM} = \frac{I_o}{2} - \frac{V_o(V_{in} - KV_o)T_s}{2V_{in}L_f} \tag{8.26}$$

If the load is light, the converter will operate in discontinuing current mode (DCM), where $i_{Lf1} + i_{Lf2}$ will reduce to zero when $v_{AB} = 0$, as shown in Figure A.1. $I_{Lf\max}$ and $I_{Lf\min}$ in DCM and the critical output current between CCM and DCM, I_G, are given in Equations 8.27–8.29, respectively, and are derived in Appendix A.

$$I_{Lf\max\_DCM} = \left(3 - \frac{4KV_o}{V_{in}}\right)\sqrt{\frac{V_{in}V_oI_oT_s}{8L_f(V_{in} - 2KV_o)}} \tag{8.27}$$

$$I_{Lf\min\_DCM} = -\sqrt{\frac{V_{in}V_oI_oT_s}{8L_f(V_{in} - 2KV_o)}} \tag{8.28}$$

$$I_G = \frac{V_o(V_{in} - 2KV_o)T_s}{2L_fV_{in}} \tag{8.29}$$

According to Equations 8.25–8.29, Figure 8.4 gives the curves of $I_{Lf\max}$ and $-I_{Lf\min}$ at different load currents under the minimum, nominal, and maximum input voltages. There is an inflexion in each curve: the corresponding current of the inflexion is the critical output current. On the left of the inflexion, the converter operates in DCM; on the right, it operates in CCM. It can be seen from Figure 8.4 that nearly all of the

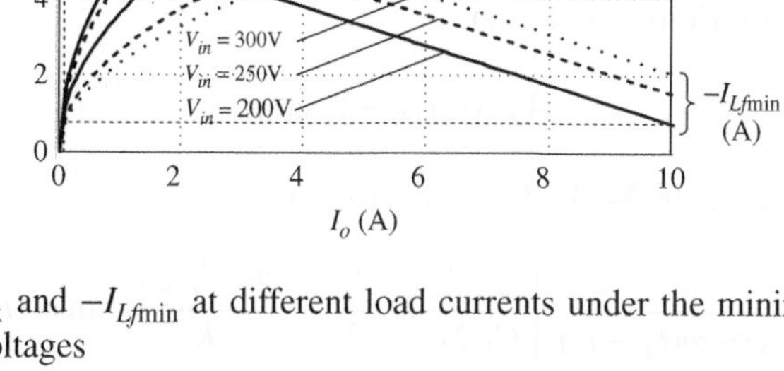

Figure 8.4 $I_{Lf\max}$ and $-I_{Lf\min}$ at different load currents under the minimum, nominal, and maximum input voltages

currents are greater than that at full load and minimum input voltage. This design of output filter inductance ensures that the lagging leg will realize ZVS at full load under minimum input voltage. Hence, both the leading and the lagging leg can realize ZVS from nearly open to full load in the input voltage range.

8.3.3 Blocking Capacitor

The blocking capacitor C_b is used to force i_p to decay rapidly when $v_{AB}=0$ so that the rectifier diodes commute naturally without voltage oscillation and the lagging leg realizes ZVS through the use of energy stored in the output filter inductor. From this point, C_b should be as small as possible. However, a small C_b will result in a high peak voltage of C_b and increased voltage stress of the rectifier diodes. Thus, the maximum C_b should be determined to ensure the commutation of the two rectifier diodes.

As shown in Figure 8.1b, the rectifier diodes finish commutation at t_3. The worst case is that $t_3=t_4$, in which case i_p reduces to $-I_{Lf\min}/K$; that is:

$$I_p(t_4) \le -I_{Lf\min}/K \tag{8.30}$$

Substituting Equation 8.30 into Equation 8.9, and given $I_p(t_2)=I_{Lf\max}/K$, we obtain:

$$I_p(t_4) = -\frac{V_{Cb}(t_2)}{\omega L_{lk}}\sin\omega(t_4-t_2) + \frac{I_{Lf\max}}{K}\cos\omega(t_4-t_2) \le -\frac{I_{Lf\min}}{K} \tag{8.31}$$

Equation 8.31 is the condition required for the rectifier diodes to finish commutation at t_4. It depends on $V_{cb}(t_2)$.

According to Figure 8.1b, the expression of $V_{cb}(t_2)$ is derived as:

$$\begin{aligned} V_{Cb}(t_2) &= V_{Cb}(t_0) + \frac{1}{C_b}\int_{t_0}^{t_2} i_p(t)dt = V_{Cb}(t_0) + \frac{1}{C_b}\cdot\frac{I_o}{2K}\cdot(t_2-t_0) \\ &= V_{Cb}(t_0) + \frac{1}{C_b}\cdot\frac{I_o}{2K}\cdot D_y\frac{T_s}{2} \end{aligned} \tag{8.32}$$

From Equation 8.10, we can obtain:

$$V_{Cb}(t_4) = \omega L_{lk} I_p(t_2)\sin\omega(t_4-t_2) + V_{Cb}(t_2)\cos\omega(t_4-t_2) \tag{8.33}$$

It can be seen from Figure 8.1b that:

$$V_{Cb}(t_4) = -V_{Cb}(t_0) \tag{8.34}$$

From Equations 8.32–8.34, $V_{cb}(t_2)$ is derived as:

$$V_{Cb}(t_2) = \frac{1}{1+\cos\omega(t_4-t_2)}\left[\frac{1}{C_b}\frac{I_o}{2K}D_y\frac{T_s}{2} - \frac{\omega L_{lk} I_{Lf\max}}{K}\sin\omega\,(t_4-t_2)\right] \tag{8.35}$$

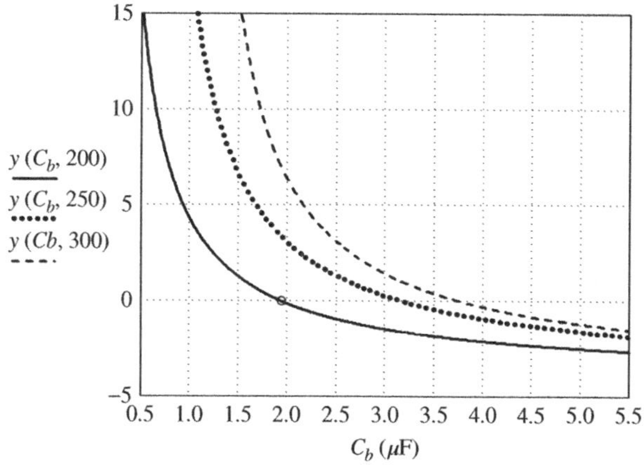

Figure 8.5 Plot of the left side of inequality Equation 8.37 versus C_b under different input voltages

where:

$$t_4 - t_2 = \frac{T_s}{2}(1 - D_y) \tag{8.36}$$

Substitution of Equations 8.16–8.18, 8.35, and 8.36 into Equation 8.31 leads to:

$$y(C_b, V_{in}) = \frac{\frac{2KV_o}{V_{in}} T_s}{\sqrt{L_{lk} C_b}} \tan \frac{\left(1 - \frac{2KV_o}{V_{in}}\right) T_s}{4\sqrt{L_{lk} C_b}} - 4 \geq 0 \tag{8.37}$$

Figure 8.5 shows the plot of the left side of inequality Equation 8.37 versus C_b under different input voltages, illustrating that in order to satisfy Equation 8.37, C_b should be smaller than 1.9 μF at the lowest input voltage $V_{inmin} = 200$ V. This is because at the lowest input voltage, the duty cycle is at its maximum and the time it takes for i_p to decay is at its shortest. We choose $C_b = 1.5$ μF.

8.4 Experimental Verification

Experimental measurements from a 540 W output power (54 V, 10 A) prototype converter are used to verify the operating principle of the ZVS PWM full-bridge converter with CDR.

The key specifications are as follows:

- input voltage $V_{in} = 250$ Vdc ± 20%;
- output voltage $V_o = 54$ Vdc; and
- output current $I_o = 10$ A.

The power devices and components for the prototype are:

- primary-to-secondary-windings-turns ratio $K = 1.5$;
- leakage inductance measured at the switching frequency $L_{lk} = 0.46\,\mu\text{H}$;
- blocking capacitor $C_b = 1.5\,\mu\text{F}$;
- output filter inductances $L_{f1} = L_{f2} = 28\,\mu\text{H}$;
- output filter capacitor $C_f = 6600\,\mu\text{F}$;
- power switches Q_1 to Q_4: IRF450;
- rectifier diodes: DSEI12-06A; and
- switching frequency $f_s = 100\,\text{kHz}$.

Figure 8.6 shows the experimental results at full load. Figure 8.6a gives the primary voltage v_{AB}, the primary current i_p, the blocking capacitor voltage v_{Cb}, and the currents

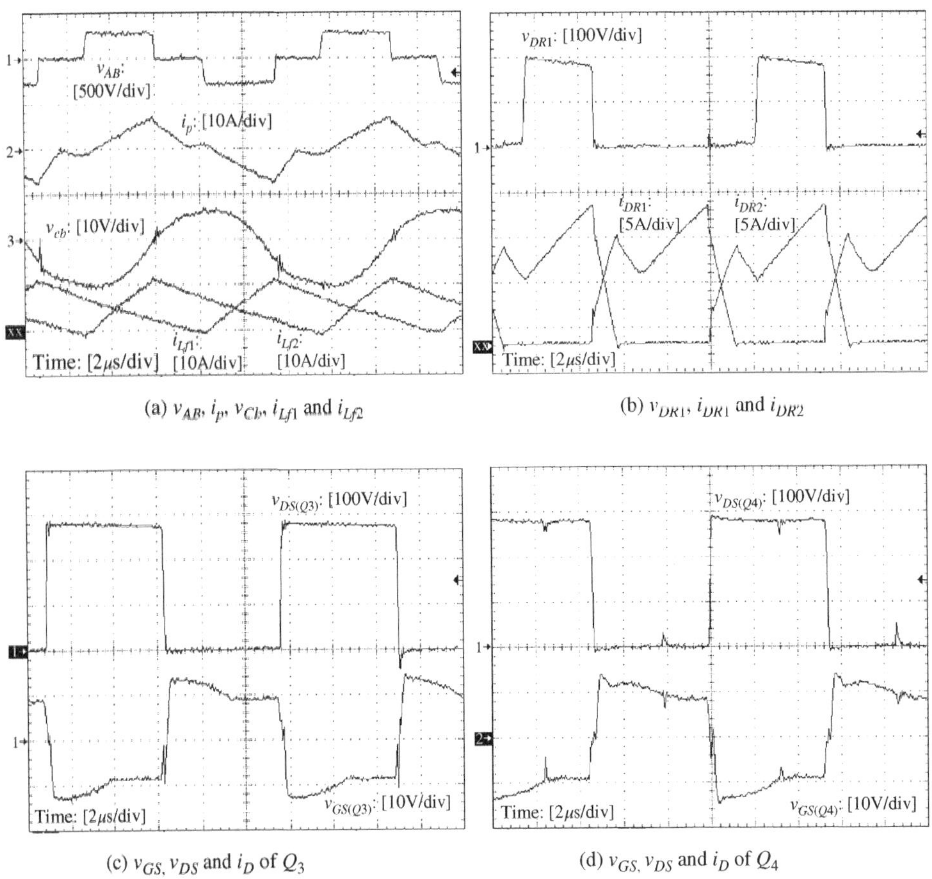

(a) v_{AB}, i_p, v_{Cb}, i_{Lf1} and i_{Lf2}

(b) v_{DR1}, i_{DR1} and i_{DR2}

(c) v_{GS}, v_{DS} and i_D of Q_3

(d) v_{GS}, v_{DS} and i_D of Q_4

Figure 8.6 Experimental results at full load of 10 A

of the two filter inductors i_{Lf1} and i_{Lf2}, showing that when $v_{AB}=0$, v_{Cb} forces i_p to decay rapidly to make the rectifier diodes commutate naturally. Figure 8.6b shows the voltage of the rectifier diodes v_{DR1} and the rectifier diode currents i_{DR1} and i_{DR2}. The rectifier diodes finish commutation prior to switching v_{AB} from zero state to active mode, and no oscillation or voltage spike occurs. Figure 8.6c,d shows the gate drive signal, the voltage across the drain and source, and the drain current of the leading-leg switch Q_3 and the lagging-leg switch Q_4. It can be seen that both the leading-leg and the lagging-leg switch realize ZVS.

Figures 8.7 and 8.8 show the experimental results at half load (5 A) and light load (1 A), respectively. At light load of 1 A, the converter operates in DCM and all the power switches realize ZVS. The natural commutation of the output rectifier diodes is also achieved and no voltage oscillation occurs.

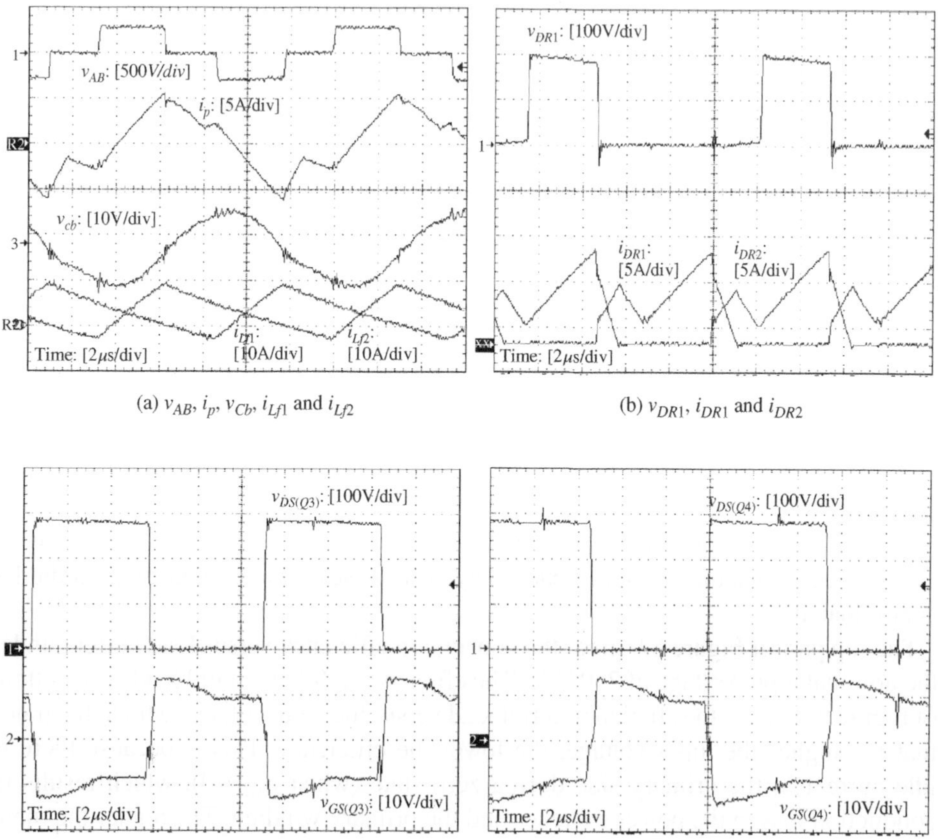

(a) v_{AB}, i_p, v_{Cb}, i_{Lf1} and i_{Lf2} (b) v_{DR1}, i_{DR1} and i_{DR2}

(c) v_{GS}, v_{DS} and i_D of Q_3 (d) v_{GS}, v_{DS} and i_D of Q_4

Figure 8.7 Experimental results at half load of 5 A

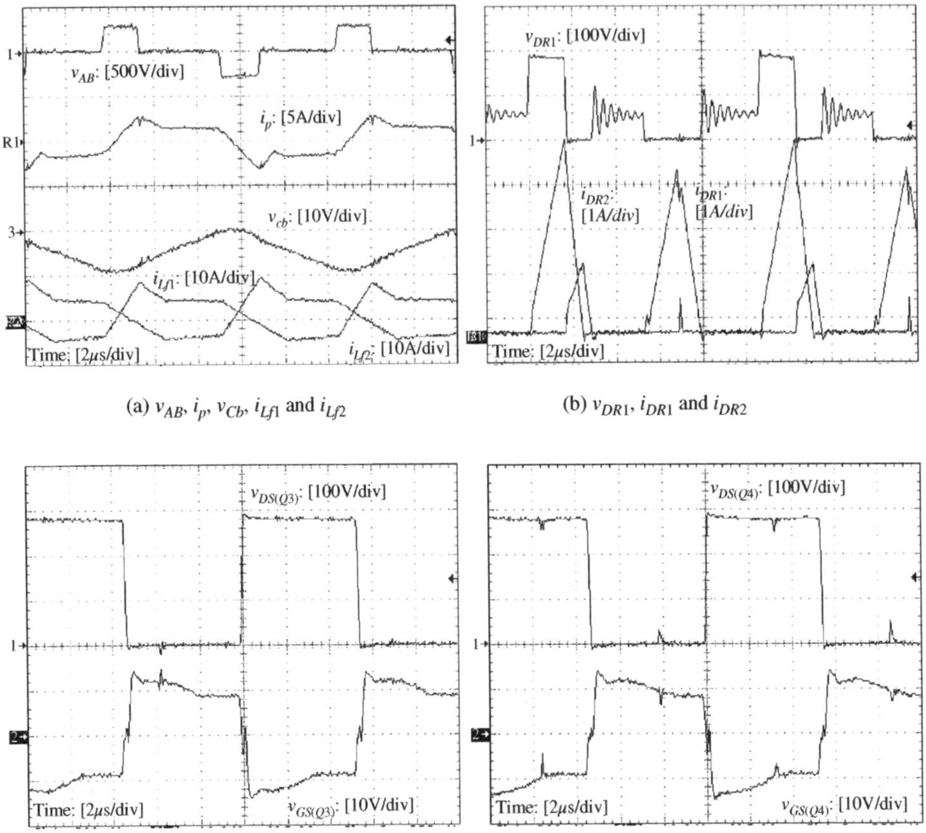

(a) v_{AB}, i_p, v_{Cb}, i_{Lf1} and i_{Lf2}

(b) v_{DR1}, i_{DR1} and i_{DR2}

(c) v_{GS}, v_{DS} and i_D of Q_3

(d) v_{GS}, v_{DS} and i_D of Q_4

Figure 8.8 Experimental results at light load of 1 A

From Figures 8.6–8.8, we know that the power switches can realize ZVS in a wide load range and that the oscillation caused by the reverse recovery of the rectifier diodes is eliminated.

The top part of Figure 8.9 gives the conversion efficiency versus load current under the nominal input voltage of 250 Vdc. The efficiency is 92.4% at full load. The bottom of Figure 8.9 shows the efficiency at full load versus the input voltage, which illustrates that the higher the input voltage, the lower the efficiency. This is because there is idle current in the primary side during zero state (when $v_{AB} = 0$), which results in conduction loss in the power switches and the primary winding. The higher the input voltage, the longer the zero state; therefore, the higher the conduction loss, the lower the efficiency.

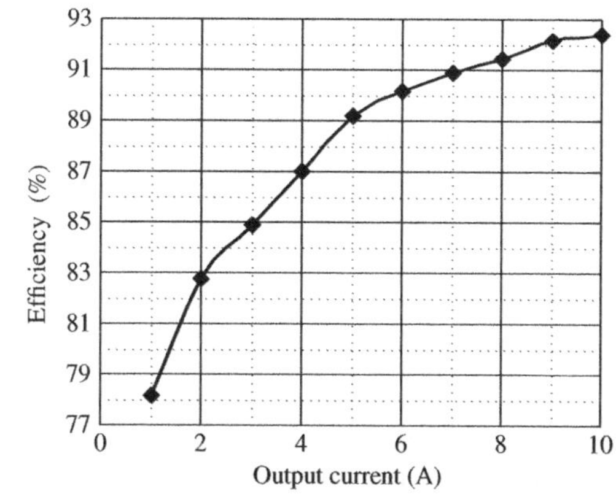

(a) Efficiency versus load current under the nominal input voltage

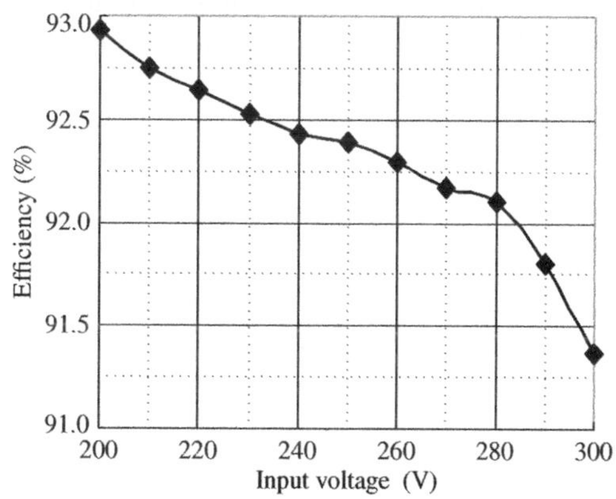

(b) Efficiency at full load versus input voltage

Figure 8.9 Measured efficiency of the CDR ZVS PWM full-bridge converter

8.5 Summary

This chapter described an improved ZVS PWM full-bridge converter with CDR by introducing a blocking capacitor in series with the primary winding of the transformer. The improved converter keeps the following advantages of the original:

1. The switches realize ZVS in a wide load range through the use of the energy of the filter inductors.
2. The rectifier diodes commute naturally without voltage oscillation or voltage spike.
3. There is no duty cycle loss, due to very small leakage inductance.

In addition, the introduction of the blocking capacitor voltage forces the primary current to decay rapidly, and it is relatively higher than the conduction voltage drop of the power switches. Hence, the transformer leakage inductor is no longer restricted, and no special manufacturer technology for the transformer is required.

The operating principle of the improved converter was analyzed, and the determination of the output filter inductance and the blocking capacitor in order to achieve ZVS was discussed in detail. The theoretical analysis was verified by a 540 W prototype converter and the experimental results were included.

References

1. Kutkut, N.H., Divan, D.M. and Gascoigne, R.W. (1995) An improved full-bridge zero-voltage switching PWM converter using a two-inductor rectifier. *IEEE Trans. Ind. Appl.*, **31** (1), 119–126.
2. Ruan, X., Wang, J. and Chen, Q. (2001) An improved current-doubler-rectifier ZVS PWM full-bridge converter. Proceedings of the IEEE Power Electronics Specialists Conference, pp. 1749–1754.
3. Ruan, X. and Wang, J. (2004) Calculation of the resonant capacitor of the improved current-doubler-rectifier ZVS PWM full-bridge converter. *IEEE Trans. Ind. Electron.*, **51** (2), 518–520.
4. Erickson, R.W. and Maksimovic, D. (2001) *Fundamentals of Power Electronics*, 2nd edn, Kluwer Academic Publishers, Boston, MA.

Appendix

Maximum and Minimum Values of the Filter Inductor Current when a CDR Full-Bridge Converter Operates in DCM and The Value of the Critical Output Current at the DCM Boundary

This appendix is provided to derive: (i) $I_{Lf\min\_DCM}$ and $I_{Lf\max\_DCM}$, the maximum and minimum values, respectively, of the filter inductance current when a current-doubler-rectifier (CDR) zero-voltage-switching (ZVS) pulse-width modulation (PWM) full-bridge converter operates in discontinuous current mode (DCM); and (ii) I_G, the critical output current at the DCM boundary.

When the output current decreases, the sum of the two filter inductor currents will reduce to zero when $v_{AB}=0$. The load is supplied by the output filter capacitor, so the CDR ZVS PWM full-bridge converter operates in DCM. The key waveforms are shown in Figure A.1.

During $[t_0, t_2]$, the two output filter inductor currents are expressed as:

$$i_{Lf1}(t) = I_{Lf\min\_DCM} + \frac{\frac{V_{in}}{K} - V_o}{L_f}(t - t_0) \tag{A1}$$

$$i_{Lf2}(t) = -I_{Lf\min\_DCM} - \frac{V_o}{L_f}(t - t_0) \tag{A2}$$

Note that here $I_{Lf\min\_DCM}$ is a negative value.

Soft-Switching PWM Full-Bridge Converters: Topologies, Control, and Design, First Edition. Xinbo Ruan.

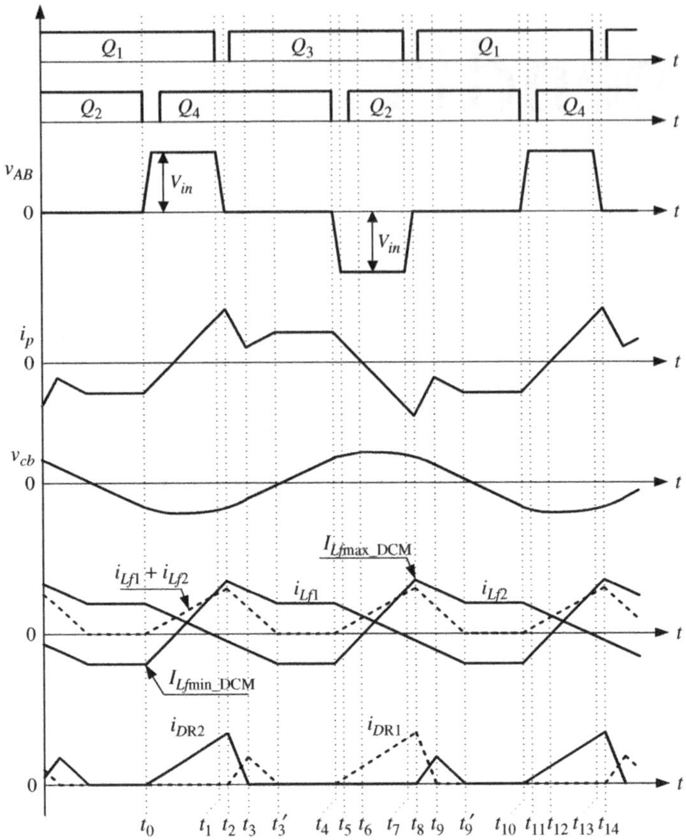

Figure A.1 Key waveforms of the CDR ZVS PWM full-bridge converter in DCM

According to Equations A1 and A2, at t_2 we have:

$$I_{Lf1}(t_2) + I_{Lf2}(t_2) = \frac{\frac{V_{in}}{K} - 2V_o}{L_f} \cdot (t_2 - t_0) \tag{A3}$$

$$I_{Lf1}(t_2) = I_{Lf\max\_DCM} = I_{Lf\min\_DCM} + \frac{\frac{V_{in}}{K} - V_o}{L_f}(t_2 - t_0) \tag{A4}$$

where:

$$t_2 - t_0 = D_y T_s/2 \tag{A5}$$

During $[t_2, t_3']$:

$$i_{Lf1}(t) = I_{Lf1}(t_2) - \frac{V_o}{L_f}(t - t_2) \tag{A6}$$

$$i_{Lf2}(t) = I_{Lf2}(t_2) - \frac{V_o}{L_f}(t - t_2) \quad \text{(A7)}$$

$$i_{Lf1}(t) + i_{Lf2}(t) = I_{Lf1}(t_2) + I_{Lf2}(t_2) - \frac{2V_o}{L_f}(t - t_2) \quad \text{(A8)}$$

At t_3', $i_{Lf1} = -I_{Lf\min\_DCM}$, $i_{Lf2} = I_{Lf\min\_DCM}$, and $i_{Lf1} + i_{Lf2}$ reduce to zero, so according to Equation A8 we have:

$$t_3' - t_2 = \frac{L_f[I_{Lf1}(t_2) + I_{Lf2}(t_2)]}{2V_o} \quad \text{(A9)}$$

Substituting Equations A3 and A5 into Equation A9 leads to:

$$t_3' - t_2 = \frac{\left(\frac{V_{in}}{K} - 2V_o\right) \cdot D_y T_s}{4V_o} \quad \text{(A10)}$$

According to Equations A2 and A7, the value of i_{Lf2} at t_3' is obtained as:

$$I_{Lf2}(t_3') = -I_{Lf\min\_DCM} - \frac{V_o}{L_f}\left[(t_2 - t_0) + (t_3' - t_2)\right] = I_{Lf\min\_DCM} \quad \text{(A11)}$$

Substituting Equations A5 and A10 into Equation A11 yields:

$$I_{Lf\min\_DCM} = -\frac{V_{in} D_y T_s}{8KL_f} \quad \text{(A12)}$$

Substituting Equations A5 and A12 into Equation A4 leads to:

$$I_{Lf\max\_DCM} = \frac{3V_{in} - 4KV_o}{8KL_f} D_y T_s \quad \text{(A13)}$$

The output current is the average value of the sum of the two filter inductor currents, that is:

$$I_o = \overline{i_{Lf1} + i_{Lf2}} = \frac{I_{Lf1}(t_2) + I_{Lf2}(t_2)}{2} \cdot \left[(t_3' - t_2) + (t_2 - t_0)\right] \Big/ \frac{T_s}{2} \quad \text{(A14)}$$

From Equations A3, A5, A11, and A14, the duty cycle in DCM is derived as:

$$D_y = \sqrt{\frac{8V_o I_o L_f}{\left(\frac{V_{in}^2}{K^2} - \frac{2V_{in}V_o}{K}\right) T_s}} \quad \text{(A15)}$$

Substituting Equation A15 into Equation A12 and A13, respectively, yields:

$$I_{Lf\min\_DCM} = -\sqrt{\frac{V_{in}V_o I_o T_s}{8L_f(V_{in} - 2KV_o)}} \tag{A16}$$

$$I_{Lf\max\_DCM} = \left(3 - \frac{4KV_o}{V_{in}}\right) \cdot \sqrt{\frac{V_{in}V_o I_o T_s}{8L_f(V_{in} - 2KV_o)}} \tag{A17}$$

If $t_3' = t_4$ then the converter operates at the DCM boundary, $t_3' - t_0 = T_s/2$, and $K = D_y V_{in}/(2V_o)$, so the critical output current I_G can be derived from Equations A3, A5, and A14 as:

$$I_G = \frac{V_o(V_{in} - 2KV_o)T_s}{2L_f V_{in}} \tag{A18}$$

Bibliography

Barbi, I., Bolacell, J.C., Martins, D.C. and Libano, F.B. (1990) Buck quasi-resonant converter operating at constant frequency: analysis, design and experimentation. *IEEE Trans. Power Electron.*, **5** (3), 276–283.

Borage, M., Tiwari, S., Bhardwaj, S. and Kotaiah, S. (2008) A full-bridge dc–dc converter with zero-voltage-switching over the entire conversion range. *IEEE Trans. Power Electron.*, **23** (4), 1743–1750.

Borage, M., Tiwari, S. and Kotaiah, S. (2005) A passive auxiliary circuit achieves zero-voltage-switching in full-bridge converter over entire conversion range. *IEEE Trans. Power Electron. Lett.*, **3** (4), 141–143.

Chen, Z., Ji, B., Ji, F. and Shi, L. (2010) Analysis and design considerations of an improved ZVS full-bridge dc-dc converter. Proceedings of the IEEE Applied Power Electronics Conference and Exposition, pp. 1471–1476.

Chen, Q., Loft, A.W. and Lee, F.C. (1991) Design trade-offs in 5-V output off-line zero-voltage, PWM converter. Proceedings of the IEEE International Telecommunications Energy Conference, pp. 616–623.

Chen, W., Ruan, X., Chen, Q. and Ge, J. (2010) Zero-voltage-switching PWM full-bridge converter employing auxiliary transformer to reset the clamping diode current. *IEEE Trans. Power Electron.*, **25** (5), 1149–1162.

Chen, W., Ruan, X. and Zhang, R. (2008) A novel zero-voltage-switching PWM full bridge converter. *IEEE Trans. Power Electron.*, **23** (2), 793–801.

Chen, Q., Yin, L., Wang, J. *et al.* (2008) Phase-shifted full-bridge PWM converter with clamping diodes and current transformer. Proceedings of the IEEE Power Electronics Specialists Conference, pp. 2403–2409.

Cho, J.G., Baek, J.W., Jeong, C.Y. *et al.* (1998) Novel zero-voltage and zero-current-switching (ZVZCS) full bridge PWM converter using a simple auxiliary circuit. Proceedings of the IEEE Applied Power Electronics Conference and Exposition, pp. 834–839.

Cho, J.G., Baek, J.W., Yoo, D.W. and Lee, H.S. (1997) Novel zero-voltage and zero-current-switching (ZVZCS) full bridge PWM converter using transformer auxiliary winding. Proceedings of the IEEE Power Electronics Specialists Conference, pp. 227–232.

Cho, J.G., Rim, G.H. and Lee, F.C. (1996) Zero voltage and zero current switching full bridge PWM converter with secondary active clamp. Proceedings of the IEEE Power Electronics Specialists Conference, pp. 657–663.

Soft-Switching PWM Full-Bridge Converters: Topologies, Control, and Design, First Edition. Xinbo Ruan.

Cho, J.G., Sabate, J.A., Hua, G. and Lee, F.C. (1994) Zero-voltage and zero-current-switching full-bridge PWM converter for high power applications. Proceedings of the IEEE Power Electronics Specialists Conference, pp. 102–108.

Cho, J.G., Sabate, J.A. and Lee, F.C. (1994) Novel full-bridge zero-voltage-transition PWM dc/dc converter for high power applications. Proceedings of the IEEE Applied Power Electronics Conference and Exposition, pp. 143–149.

Divan, D.M. (1989) The resonant dc link converter – A new concept in static power conversion. *IEEE Trans. Ind. Appl.*, **25** (2), 317–325.

Erickson, R.W. and Maksimovic, D. (2001) *Fundamentals of Power Electronics*, 2nd edn, Kluwer, Boston, MA.

Fisher, R.A. and Ngo, K.D.T. (1988) A 500 kHz, 250 W dc-dc converter with multiple outputs controlled by phase-shifted PWM and magnetic amplifiers. Proceedings of the High Frequency Power Conversion Conference, pp. 100–110.

Hamada, S. and Nakaoka, M. (1994) Analysis and design of a saturable reactor assisted soft-switching full-bridge dc-dc converter. *IEEE Trans. Power Electron.*, **9** (3), 309–317.

Hua, G. and Lee, F.C. (1991) A new class of zero-voltage-switched PWM converters. Proceedings of the High Frequency Power Conversion Conference, pp. 244–251.

Hua, G., Lee, F.C. and Jovanovic, M.M. (1991) An improved zero-voltage-switched PWM converter using a saturable inductor. Proceedings of the IEEE Power Electronics Specialists Conference, pp. 189–194.

Hua, G., Leu, C.S., Jiang, Y.M. and Lee, F.C. (1992) Novel zero-voltage-transition PWM converters. Proceedings of the IEEE Power Electronics Specialists Conference, pp. 55–61.

Hua, G., Yang, E.X., Jiang, Y.M. and Lee, F.C. (1993) Novel zero-current-transition PWM converters. Proceedings of the IEEE Power Electronics Specialists Conference, pp. 538–544.

Jang, Y. and Jovanovic, M.M. (2004) A new family of full-bridge ZVS converters. *IEEE Trans. Power Electron.*, **19** (3), 701–708.

Jang, Y. and Jovanovic, M.M. (2007) A new PWM ZVS full-bridge converter. *IEEE Trans. Power Electron.*, **22** (3), 987–994.

Jang, Y., Jovanovic, M.M. and Chang, Y.M. (2003) A new ZVS-PWM full-bridge converter. *IEEE Trans. Power Electron.*, **18** (5), 1122–1129.

Kim, E.S., Joe, K.Y., Kye, M.H. *et al.* (1998) An improved ZVZCS PWM full-bridge dc/dc converter using energy recovery snubber. Proceedings of the IEEE Applied Power Electronics Conference and Exposition, pp. 1014–1019.

Kim, I.D., Nho, E.C. and Cho, G.H. (1992) Novel constant frequency PWM dc/dc converter with zero-voltage-switching for both primary switches and secondary rectifying diodes. *IEEE Trans. Ind. Electron.*, **39** (5), 444–452.

Kutkut, N.H., Divan, D.M. and Gascoigne, R.W. (1995) An improved full-bridge zero-voltage switching PWM converter using a two-inductor rectifier. *IEEE Trans. Ind. Appl.*, **31** (1), 119–126.

Lee, F.C. (1988) High-frequency quasi-resonant and multi-resonant converter technologies. Proceedings of the IEEE International Conference on Industrial Electronics, pp. 509–521.

Lofti, A.W., Sabate, J.A. and Lee, F.C. (1990) Design optimization of the zero-voltage-switched PWM converter. Proceedings of the Virginia Power Electronics Center Seminar, pp. 30–37.

Mweece, L.H., Wright, C.A. and Schlecht, M.F. (1989) A 1 kW 500 kHz front-end converter for a distributed power supply system. Proceedings of the IEEE Applied Power Electronics Conference and Exposition, pp. 423–432.

Oruganti, R. and Lee, F.C. (1985) Resonant power processors, Part I: State plane analysis. *IEEE Trans. Ind. Appl.*, **21** (6), 1453–1461.

Oruganti, R. and Lee, F.C. (1985) State-plane analysis of parallel resonant converters. Proceedings of the IEEE Power Electronics Specialists Conference, pp. 56–73.

Patterson, O.D. and Divan, D.M. (1987) Pseudo-resonant full bridge dc/dc converter. Proceedings of the IEEE Power Electronics Specialists Conference, pp. 424–430.

Pressman, A.I., Billings, K. and Morey, T. (2009) *Switching Power Supply Design*, 2nd edn, McGraw-Hill Professional, New York.

Redl, R., Balogh, L. and Edwards, D.W. (1993) Switch transitions in the soft-switching full-bridge PWM phase-shift dc/dc converter: analysis and improvements. Proceedings of the International Telecommunication Energy Conference, pp. 350–357.

Redl, R., Balogh, L. and Edwards, D.W. (1994) Optimum ZVS full-bridge dc/dc converter with PWM phase-shifted control: analysis, design considerations, and experimental results. Proceedings of the IEEE Applied Power Electronics Conference and Exposition, pp. 159–165.

Redl, R., Sokal, N.O. and Balogh, L. (1991) A novel soft-switching full-bridge dc/dc converter: analysis, design considerations and experimental results at 1.5 kW, 100 kHz. *IEEE Trans. Power Electron.*, **26** (3), 408–418.

Ruan, X. and Liu, F. (2004) An improved ZVS PWM full-bridge converter with clamping diodes. Proceedings of the IEEE Power Electronics Specialists Conference, pp. 1476–1481.

Ruan, X. and Wang, J. (2004) Calculation of the resonant capacitor of the improved current-doubler-rectifier ZVS PWM full-bridge converter. *IEEE Trans. Ind. Electron.*, **51** (2), 518–520.

Ruan, X., Wang, J. and Chen, Q. (2001) An improved current-doubler-rectifier ZVS PWM full-bridge converter. Proceedings of the IEEE Power Electronics Specialists Conference, pp. 1749–1754.

Ruan, X. and Yan, Y. (1996) Zero-voltage-switched, PWM, phase-shifted converter employing an auxiliary resonant net. Proceedings of the IEE Power Electronics and Variable Speed Drives, pp. 534–539.

Ruan, X. and Yan, Y. (2000) Soft-switching techniques for PWM dc/dc full-bridge converters. Proceedings of the IEEE Power Electronics Specialists Conference, pp. 634–639.

Ruan, X. and Yan, Y. (2001) A novel zero-voltage and zero-current-switching PWM full bridge converters using two diodes in series with the lagging leg. *IEEE Trans. Ind. Electron.*, **48** (4), 777–785.

Sabate, J.A., Vlatkovic, V., Ridley, R.B. *et al.* (1990) Design considerations for high-voltage, high power full-bridge zero-voltage-switched PWM converter. Proceedings of the IEEE Applied Power Electronics Conference and Expostion, pp. 275–284.

Sabate, J.A., Vlatkovic, V., Ridley, R.B. and Lee, F.C. (1991) High-voltage, high power, ZVS, full-bridge PWM converter employing an active snubber. Proceedings of the Virginia Power Electronics Center Seminar, pp. 125–130.

Sable, D.M. and Lee, F.C. (1989) The operation of a full-bridge, zero-voltage-switched PWM converter. Proceedings of the Virginia Power Electronics Center Seminar, pp. 92–97.

Spiazzi, G., Buso, S., Citron, M. *et al.* (2003) Performance evaluation of a Schottky SiC power diode in a boost PFC application. *IEEE Trans. Power Electron.*, **18** (6), 1249–1253.

Tabisz, W.A. and Lee, F.C. (1988) Zero-voltage-switching multi-resonant technique – A novel approach to improve performance of high frequency quasi-resonant converters. Proceedings of the IEEE Power Electronics Specialists Conference, pp. 9–17.

Xu, M. (1997) Research on PWM soft-switching topology theory. PhD dissertation. Zhejiang University, Hangzhou, China.

Yang, B., Duarte, J.L., Li, W. *et al.* (2010) Phase-shifted full bridge converter featuring ZVS over the full load range. Proceedings of the Annual Conference of the IEEE Industrial Electronics Society, pp. 644–649.

Yin, L., Chen, Q., Peng, B. *et al.* (2007) Key issues of clamping diodes in DCM phase-shift full-bridge converter. Proceedings of the IEEE Power Electronics Specialists Conference, pp. 1721–1725.

Zhang, Q., Callanan, R., Das, M.K. *et al.* (2010) SiC power devices for microgrids. *IEEE Trans. Power Electron.*, **25** (12), 2889–2896.

Index

Soft-Switching PWM Full-Bridge Converters: Topologies, Control, and Design, First Edition. Xinbo Ruan.

Printed and bound by CPI Group (UK) Ltd, Croydon, CR0 4YY
21/09/2026
14966532-0001